FRIEDRICH RATZEL

FRIEDRICH RATZEL

A BIOGRAPHICAL MEMOIR
AND BIBLIOGRAPHY

BY

HARRIET WANKLYN
(MRS J. A. STEERS)

*Lecturer in Geography in the
University of Cambridge*

CAMBRIDGE
AT THE UNIVERSITY PRESS
1961

PUBLISHED BY
THE SYNDICS OF THE CAMBRIDGE UNIVERSITY PRESS

Bentley House, 200 Euston Road, London, N.W.1
American Branch: 32 East 57th Street, New York 22, N.Y.
West African Office: P.O. Box 33, Ibadan, Nigeria

©

CAMBRIDGE UNIVERSITY PRESS

1961

Printed in Great Britain at the University Press, Cambridge
(Brooke Crutchley, University Printer)

For
A.M.B.

Who taught us the beginnings of
German and so much else besides

CONTENTS

Preface *page* ix

Part I

A BIOGRAPHICAL MEMOIR 1

NOTES 46

Part II

A BIBLIOGRAPHY OF RATZEL'S WORK 55

  I Books 57

  II Papers and shorter contributions to journals and collective works 61

Index 95

PREFACE

THE material for this short biographical essay has been collected from scattered sources. The work connected with it has mainly been that of seeking out and assembling an outline account of a geographical professor's life, no part of which is really unknown, but which does not seem to have appeared before in coherent and accessible form to English readers.

For present-day students of Ratzel, by far the most valuable book is that by Dr Steinmetzler, *Die Anthropogeographie Friedrich Ratzels und ihre ideengeschichtlichen Wurzeln*, published at Bonn in 1956, to which frequent allusions are made in this essay. Dr Steinmetzler had access to the manuscripts of many of Ratzel's lectures and of some of his written works: he was also able to examine the scanty remains of Ratzel's correspondence. The result is a most useful and compact analysis of the character of Ratzel's anthropogeography, and on this most geographers must lean heavily in their own studies. His work also makes known the studies of Ratzel by O. Schluter at the beginning of the century and the more modern ones by H. Schmitthenner.

I am most grateful to him and to Professor Troll of the Geographical Institute, Bonn University, for help with this work at various stages, and to Mr Hermann Franke of the Institute of Theoretical Physics, Bonn University; to Professor Jan Broek of the Department of Geography, the University of Minnesota, to Dr Sydney Smith of St

PREFACE

Catharine's College, Cambridge and to Mr G. I. Jones of the Department of Anthropology, Cambridge University, for reading and correcting the typescript; to Professor Lehmann of the Deutsches Institut für Länderkunde in Leipzig for arranging the loan of the *Bericht des 'Geographischen Abends'* (1901) and of Eckert's study of Friedrich Ratzel, published in 1927, since copies of these are now hard to come by; to the Librarian of the University Library at Cambridge for his good offices in facilitating this loan from East Germany; to Miss Felland of the Library of the American Geographical Society, and to Dr Alastair Crombie of All Souls College, Oxford, for suggestions as to sources for the raw material for this essay; to Miss Jaqueline Bull, Archivist of the Margaret I. King Library, Lexington, University of Kentucky, for thermofax copies of newspaper reports referring to Ellen Semple's studies in Leipzig; to Dr Hans Weigert of Georgetown University for obtaining a microfilm copy of Dr Parrella's thesis, *Lebensraum and Manifest Destiny*; to Dr Erika Sellenberg of Bonn, Miss Morris of Newnham College, and to Mrs Stopp for help with the translation of some titles in the bibliography, and last, but not least, to the Trustees of the British Museum Library, for permission to study in the Reading Room the considerable collection of Ratzel's works, and to the photographic department for producing the photostat copy of Hantzsch's bibliography without which an essay of this kind would have been impossible.

H. G. WANKLYN

Cambridge, 1960

PART I

A BIOGRAPHICAL MEMOIR

I

RATZEL'S identity and reputation as a geographer are slowly returning to reality. It is interesting to compare the obituary notices of this scholar and his work, some critical, but most of them affectionate and respectful, published throughout Europe and America in 1905 and 1906, with the near-phobia with which some of his ideas have been cited in the last twenty years,[1] and then to turn again to the emerging recognition of this German geographer in quite recent writing.[2]

One realization from examining this sharply varying opinion is that of the limited study of Ratzel's written work. Certain titles have become notorious: the two volumes of *Anthropogeography* (see the Bibliography in the second half of this book, Part I, 9 and 14); the great treatise on political geography (Part I, 18); the three-volume study of the peoples of the world in the original and English versions (Part I, 11 and 17); the relatively short treatise on the influence of sea power (Part I, 22), and above all the eighty-page article on 'Der Lebensraum' (Part II, 295) written near the end of his life as part of a complimentary volume presented to Albert Schäffle. Too often, unfortunately, modern geographers have seized upon passages and even sentences from Ratzel's writings without appreciating the context. They have not always realized the changing phases of the author's thought and work, zoological, geographical and philosophical, and some of Ratzel's books and papers have lent themselves very readily to misunderstanding and perversion in a world of distracted political thinking. It would indeed be possible to use or misuse short passages of his written work to support almost any wild-goose chase in geographical thought.

It needs a considerable effort to read Ratzel's political geography and many of his shorter articles, and to rid the mind of the role of the Germans in Europe and overseas in the last fifty years. But the failure to do so, and to explore the whole range of his work in its proper setting, has been unfair to his scholarship and damaging to geography as a whole; and there are signs of conscience stirring on these points. To urge this is not to whitewash Ratzel. No one can plod through the series of articles in the *Grenzboten*,[3] for example, from 1890 onwards, without recognizing a very hearty form of polemics in political geography. But Ratzel had his integrity as a scholar and writer, and he was as different from the unpleasant geopoliticians who succeeded him as he was from the philosophical geographers who went before him.

It has thus seemed sensible to base this study upon Hantzsch's careful bibliography included in the *Kleine Schriften* in 1906 and to offer a shortened English version of it. From the bibliography there emerges at once Ratzel's enormous scale and energy as a scholar. 'Fleiß' is (or was) a German quality which has often awed and dismayed less industrious Europeans, and 'diligence' seems at times a feeble translation of the word. Ratzel's 'Fleiß' both as writer and lecturer impressed even his own contemporaries and subsequent German students of his work. His outpouring of writing certainly had its critics. Alfred Hettner,[4] who was on Ratzel's staff in Leipzig, Hermann Wagner in his reviews of Ratzel's books, and much later Hans Schrepfer[5] wrote sharply of a superficiality in thinking and writing, and of the amorphous mass of knowledge and ideas which were never disciplined into systematic thinking of lasting value. But even with these strictures, Ratzel's vigorous output of written work shows him as the product of his country and his times. It was still possible in the last quarter of the nineteenth century for a professor, especially a German professor, untrammelled by administration, welfare work and domestic chores, to acquire an extraordinary sweep of knowledge, to pronounce on his findings with confidence, and to print a good many of them with effect. And even if, in analysis half a century later, so much that he wrote and said shows the weakness of his generation of scholars, that they were content with generalization and too ready to rush in with opinion where angels fear to tread, how much greater the need to see, if only in outline, the life and work of such a man as a whole, and not just odd pieces of them.

The German geographers have not so far attracted much in the way of biographical effort from Anglo-Saxon geographers, although the centenary celebrations of Humboldt and Ritter in 1959 have brought to notice the literature about them in German.[6] Alexander von Humboldt and Karl Ritter are so outstanding that the lack of attention given to them as people by other than German geographers is surprising in a century which claims to have seen the establishment of academic geography: three others, Reinhold and Georg Forster in the eighteenth century, and Friedrich Ratzel in the nineteenth, would also repay any trouble taken to know them as people, as distinct from sporadic searches into their written work. In the case of Ratzel, as Steinmetzler has pointed out, there are difficulties in obtaining necessary raw material. Neither a diary nor much in the way of correspondence has come to light, although the short autobiography[7] found by his widow, and printed in the first volume of the *Kleine Schriften*, gives an interesting account of his early years. Many questions which arise automatically from handling the available data will probably remain unanswered: for example, what exactly were the circumstances of Ratzel's appointment to the Hochschule at Munich in 1875; and again, was there ever any personal meeting between him and Mackinder? Any student of Ratzel as a person is driven back to deduce what he can from Ratzel's writings, from contemporary descriptions of him and from the accounts of his lecturing; full-scale biography is perhaps out of the question. The sheer bulk of his writings would make the thorough reading of them the work of years and the aim of this short essay is not to rehearse again the better known and more freely discussed pieces of work. Moreover, as a German geographer, T. Achelis, was to remark after Ratzel's death, 'Wer nicht mit einem Tropfen philosophischen Öles gesalbt ist, wird sich nie und nimmer das Verständnis Ratzels erschließen'.[8] Those of us whose philosophical equipment is meagre can well understand the force of this comment and may remain baffled by much of the written work of Ratzel's last years. But there is today a need, which even a short study can help to satisfy, to see Ratzel as a person in the Germany of his time, and to recognize the range and development of his geographical scholarship through the assortment of his writings. He has been charged so long with the hatching of spurious notions of determinism and *Lebensraum* that it seems fair now to try to view the self-made scholar and the geographer at Munich and Leipzig as objectively as possible.

II

Ratzel was born at Karlsruhe in Baden on 30 August 1844. He grew up in the family of the *Kammerdiener* (manager of household staff) to the grand duke; and the first charming essays of the collection *Glücksinseln und Träume* (Part I, 26, and Part II, 336), published posthumously in 1905, give a part imaginative, part autobiographical account of his childhood on the edge of the grand-ducal household. Karl Hassert[9] in his obituary notice of Ratzel mentions the regard of the grand duke for his *Kammerdiener*, and of the absence of this much trusted servant from Baden for long spells in attendance when the grand-ducal family was travelling for education or pleasure. Ratzel and the three other members of the family (two boys and a girl), were thus brought up largely by the mother, and Ratzel's writing leaves no doubt of his deep and lasting affection for her. This relationship, together with the access to the great library in the palace and the peculiar setting of the little house within the great house, were the remembered features of his early years.

The three boys in a hard-working and intelligent family seem to have been easy to start in life. One became an architect and professor in the Technical High School at Karlsruhe and another went into business. Ratzel himself, at the age of fifteen, and after six years at the La Fontaine High School in Karlsruhe, was apprenticed to an apothecary at Eichtersheim in the Kraichgau district of Baden, between Karlsruhe and Heidelberg, and worked there for four years.

Ratzel writes himself in his short autobiography that the *Kammerdiener* and his wife had been anxious to find something in the way of apprenticeship which was more or less congenial to the boy. He had agreed with them that the opening at Eichtersheim was a good chance, and freely admitted later how much he had learned of local botany and natural history and topography through this kind of work in his teens. The recollection of the end of childhood was sharp enough, however, to enable him to write at the end of his life, and with a full allotment of German sentiment, the essay on 'Heimweh', with its description of the last meal with his parents who had brought him to Eichtersheim, the watching of their departure on the road home, and the miserable scramblings up the hillsides in the free time of the first weeks away from home to look out in the direction of Karlsruhe.

Once established, however, Ratzel seems to have worked contentedly at Eichtersheim, and later for two years in the same capacity at Rapperswyl on Lake Zürich in Switzerland (1863–5) and at Mörs near Krefeld in the Ruhr area (1865–6).

It was during the years at Rapperswyl that he seems to have entered upon a grounding in the classics. He wrote later and with gratitude of the help given him in Switzerland by Früh, the Swiss professor, and at Mörs by Breuker, in covering patchily the work which belonged properly to the last years at school.

At the age of twenty-one Ratzel had managed to persuade his family that he was intent on, and capable of, university education. He returned for a short while to the Hochschule at Karlsruhe and then began in good earnest on his zoological studies, first at Heidelberg and then at Jena and Berlin. By 1868 he had produced the thesis for his degree at Heidelberg ('Contributions to the anatomical and general study of Oligochaetes' (Part II, 3)) and a year later had published his first book *Sein und Werden der organischen Welt*. This was a thoughtful if immature comment on the significance of Darwin's work, but coming out only shortly in advance of Haeckel's much more notable *Natürliche Schöpfungsgeschichte*, it attracted little attention. Ratzel had indeed taken his views on evolution largely from Haeckel's earlier writing and made no secret at this time of his unqualified admiration for the work of the Jena zoology professor.

It is interesting at this stage, however, to remember Ratzel's hard entry from relatively humble surroundings into academic life, the speed with which he established himself, and the conviction with which he viewed the natural sciences, and zoology in particular, as offering him the academic possibilities for which he had longed. In this early zoological work, also, there is clear the whole-hearted enthusiasm for Darwin's findings which has sometimes and wrongly been cited as a permanent trait of his geographical work and interpreted as determinism.

After completing his degree, Ratzel had the chance of travel and field-work in the Mediterranean, and of working with the French naturalist, Charles Martin, at Cette and Montpellier. He was not, of course, well off, and one incident in particular, the theft of a microscope from a French railway station, roused him to the need for making some ready money. As a chance venture he offered some popular accounts of his work as a naturalist on the Mediterranean coasts to the *Kölnische Zeitung* (Part I, 2), and they were at once accepted by the paper as a novel but promising enterprise in journalism. Shortly after this success Ratzel was offered a post in the Natural History Museum at Stuttgart, but he refused it in favour of permanent work with the *Kölnische Zeitung* as a travel correspondent. For the next few years, with the important break of the Franco-Prussian War, the possibilities of travel and journalism filled the horizon for him, beginning with the considerable Italian journeys of 1869–70.

This change in Ratzel's prospects and interests is significant. It marks the move away from zoology towards one of the ingredients of geography, travel undertaken with the idea of careful observation, and shows up a fluency in writing which was to mark his work all his life. Ratzel never really ceased to be a journalist, and those who reproached him later with the charge of hack work and with the dispersion of energy and time over scores of topics, in the *Grenzboten*, the *Umschau*, the *Allgemeine Zeitung*, the

Woche, the *Gegenwart* and countless other publications, perhaps forgot the strength of this instinct in him, and the need to exercise it in both prose and verse. Contemporary and later geographers could accuse Ratzel of careless writing, of superficialities, and of inability to struggle sturdily with the mass of facts and ideas which he had collected in the course of half a lifetime. But what he wrote for the periodicals and papers was always lively, catching attention and remembrance, and in all his written work he remains one of the stylists amongst the German geographers.

The two years that followed his decision to travel and write for the *Kölnische Zeitung* were quite different, however, from what he had imagined, and also firmly moulded Ratzel's thought and work for the rest of his life. Ratzel's generation saw the formative period of German political unity, and for many of them it was a deeply moving political and emotional experience. For him, as for many others in Germany, the Franco-German crisis of 1870 was indeed the '*Einheitskrieg*' and the last stage in the prolonged struggle for national unity. In the summer of that year he volunteered for military service and enrolled in the Baden Infantry: in September he was slightly wounded at Neudorf in front of Strasbourg, and in November his military service ended with severe head injuries at the siege of Auxonne. The deep impression that the war and his active service in it made on him were apparent throughout his life. The essay called 'Thunder Heat' ('Gewitterschwüle') in *Glücksinseln und Träume* conveys vividly the excitement of so many young Germans, as the two countries came to the verge of war, and another on the military hospital ('Im

Lazarett') (Part II, 321) recalls his reactions to serious wounding and disability.

Ratzel regretted none of it and remained of the same way of thinking to the end of his life. He had considered it right to serve, and had heartily enjoyed much of army life. He had excelled as a volunteer and was proud of the decorations—the Iron Cross and the Karl Friedrich military service medal—which distinguished his short period as a combatant. Nothing angered him more than the periodic expressions of opinion in European politics condemning the war and its results. An article in the *Grenzboten* 'About our good friends the Swiss' (Part II, 167), is an example of his resentment of such criticism, in this case the outspoken ill-will of the French Swiss against the annexation of Alsace-Lorraine. Ratzel was ready enough to see the importance of Franco-German co-operation in Europe and wrote of it again in the *Grenzboten* in the following year (Part II, 187). But he was sensitive to any adverse comment on the war of 1870-1 and perhaps never saw the episode with any detachment.

It is interesting to compare the lively and almost naïve patriotism of Ratzel in the setting of the growing power of Germany in Europe with the reaction of the great geographers, Ritter and Humboldt, to the dramatic happenings of their own times. Both of them lived and worked through the upheaval of the Napoleonic wars in Europe and overseas, and through the final Prussian effort to drive the French back into France from the east. The conflict in Humboldt's career between his Parisian and Prussian affinities is one of the more interesting if less glorious aspects of his life: certainly the Prussian traditions of his family roused him to no commitments on his own part. Ritter, who was a West German from Quedlinburg, had qualms about the need to serve in the last years of the struggle against the French,[10] but decided that his first duty was still that of tutor and part-guardian to the Bethmann-Hollweg family. It is instinctive, but hardly fair, to see in Ratzel's vociferousness the beginning of the Germany which was to ruin Europe. His contemporaries at any rate had no such suspicions.[11]

III

Ratzel returned from the wars to work with the *Kölnische Zeitung*, but before beginning his professional travels and writing he had a short spell of study at Munich. There he met again K. von Zittel, the geologist, who had taught him at Karlsruhe, and what mattered more for his future, Moritz Wagner, the naturalist and curator of the Ethnographical Museum at Munich. Wagner's company and scholarship brought to the fore again Ratzel's interest in Darwin's work, although Wagner, as Darwin was to point out later, had no very accurate understanding of the Englishman's thinking.[12] Ratzel certainly absorbed from Wagner the German naturalist's own theories of the importance of the migration of species. It was perhaps this brief period at Munich which gave Ratzel his first direct awareness of the interest of geographical work in its own right, and this awareness accompanied him through the next years of very active travelling for the *Kölnische Zeitung* which took him over much of Europe and the New World. In the summer and autumn of 1871 he made his journeys through the Hapsburg Empire and wrote his accounts of the Carpathians, especially Transylvania and Bukowina; in the autumn of the same year he was in the Hungarian Plains and Budapest.

It is well to remember the Hungary of his visit. These were the years of the opening up of the Magyar lowlands to agriculture after the long stagnation which accompanied the Turkish conquest and occupation, and which persisted for generations after the Turks had withdrawn. In many parts of Central and Eastern Hungary the consciousness of security, the new assets of rail transport and steam shipping, an industrial market, especially in Germany, and here and there the intelligent farming of a great landowner made effective by countless and often conscientious estate agents, were transforming the Alföld. The Magyars indeed were beginning to appreciate, settle and use their hereditary *Lebensraum*. In the Transylvanian Highlands, to the northeast, Ratzel must have seen for himself the characteristic element of settlement in this upland basin up to the time of the Second War, the 'Saxon' predominance of long standing in the towns of Klausenburg, Hermannstadt, Kronstadt, and others. It formed the most outstanding example, perhaps, of German minority settlement in East Central Europe, and a problem which later was to have great political importance on the continent.

Accounts of these journeys, and those of 1872 in the Alps and throughout Italy, were published in the *Kölnische*

Zeitung (Part II, 15, 17, 18, 21) and many of them appeared a year later in book form, *Travels of a Naturalist* (Part I, 2).

In the two years that followed, 1874 and 1875, Ratzel went much further afield on a long tour in North and Central America, beginning his journey with a visit to London. From the New World he returned convinced of the attraction and importance of geographical work and, in his thirtieth year, certain at last of his academic vocation. The meeting at Harvard with Agassiz, the zoologist and glaciologist, was always remembered, but it was the phase of development in the United States as a whole which fascinated Ratzel and which for some time clearly influenced his descriptive writing and his philosophy of human geography.

At the time when he travelled in America, the Civil War was for most Americans a recent and humiliating memory, and from it developed the perplexity about the prospects and roles of the Africans and Asiatics in the Federation. Ratzel was sharply aware of the problem of coloured people in the States, African, Chinese and Japanese alike. The Asiatic inflow to California certainly suggested to him another and quite different theme of *Lebensraum*. These topics were written up later at Munich (see Part II, 26, 28, 63, 68, 76). What he saw of Chinese settlement in Pacific America led him to a much wider study of Chinese emigration as a whole, and it was this subject that he presented eventually as a thesis (Part I, 5) when appointed lecturer at the Munich Technical High School. One aspect of this publication is very interesting. Ratzel, in his research on Chinese settlement overseas, was brought up against the literature and practice of British colonial methods in South-

East Asia, and he used the former extensively.[13] He wrote out a clear, informed analysis of Chinese emigration in the nineteenth century, but it serves to emphasize Mackinder's reminder[14] that if German geographical scholarship at that date was the most abundant and articulate in the world, it relied very much on the data provided by British travel and exploration and on the practical findings of colonialism.

Ratzel watched the dilemma caused by the numbers of the negro population in America and by the steady inpouring of the Asians. He also studied with interest the withdrawal of the American Indian before all the newcomers, European and others. What he saw here of the shrinking of habitable areas, of a declining population, of withering primitive economies, certainly gave him his material for the ideas on the problems of marginal peoples which were later to appear in his human geography.

Finally he was immensely impressed by the huge part that his own people were playing in the opening up of the States, especially in the farming of the Middle West. By the time that Ratzel travelled in America, the effects of the great migrations of Germans in the mid-nineteenth century were apparent. He wrote of their vast numbers, their skilled farming, their social and cultural significance as a substantial peasant population squeezed out of Europe by violent social and economic change and finding its *Lebensraum* in America but, as he admitted and without rancour, with very little *political* consequence at all (Part I, 6, Vol. II, pp. 163, 593, 596–8).

Ratzel used the experiences of the American journey all his life. They appeared at length and in book form in 1878 and 1880 (Part I, 4 and 6) when he was established as a geographer at Munich. Together with his account of Mexico (Part I, 7), the American volumes rank very high both as travel books and as descriptive geographical writing, and were recognized as outstanding by geographical contemporaries like Kirchhoff in Germany and the French scholar, Réclus. Ratzel was to write again of the United States many years later when he realized their coming strategic and political importance (Part I, 15), and several of his shorter pieces of writing were concerned with them. His treatise on the sea as the source of national greatness, published in 1900 (Part I, 22), bears a strong likeness to Alfred Thayer Mahan's *Influence of Sea Power on History* which had appeared ten years earlier, and the evolution of a great political force which derived evidently from both sea and land power attracted Ratzel throughout the years.

In the earlier long works on the United States and Mexico, in his *Travels of a Naturalist* and in his study of Chinese emigration, Ratzel showed in his late twenties a power of vivid descriptive writing which in his shorter efforts never wholly left him. These books of the eighteen-seventies recall a little the clear vivacity of Humboldt's description of Mexico and his essay on the Isle of Cuba.[15] Both sets of early books are readable today, when the more profound treatises of both scholars, Humboldt's *Kosmos* and Ratzel's *Political Geography*, make the modern geographer yawn. It is not only that he now prefers information or opinion to be compressed within tighter limits than seven or eight hundred pages. Both men wrote at length whatever their theme. But from the point of view of the Anglo-Saxon reader there seem to have descended so often upon German geographers in maturity an earnestness of purpose, a verbosity and a tedious inflow of philosophical reflection which make it hard to believe that the same person could earlier have written descriptions which remain classics of geographical literature.

Another likeness between Humboldt's and Ratzel's activities appears here. Humboldt's reputation was made as a traveller, and his journeyings in the New World were indeed spectacular. But his years of travel, even adding those of his European field-work, were very few in relation to an immensely long life, and the intellectual energy which marked all of it certainly derived only in part from his time in the Americas. Ratzel also travelled extensively (though in a relatively easy way) and for a few years only, and his output as a scholar and writer again rested only to a limited extent on his own observations. In a much humbler fashion, and apparently from a sense of duty rather than enjoyment, Ritter, who is often termed an armchair geographer, worked more according to present-day relationships between travel and teaching, making such carefully planned journeys every summer as his means allowed of, and thus becoming acquainted with the greater part of Europe.[16]

When he came back from America in 1875, Ratzel resigned from his post with the *Kölnische Zeitung* to accept an appointment as lecturer in geography to the Technical High School at Munich. After a year he was offered an assistant professorship there, and in 1880 the full professorship. Two other offers were made to him at this time; one was the direction of Justus Perthes's firm at Gotha, and the other the editorship of the periodical *Ausland*. Ratzel, however, was now set in his course as an academic geographer, and remained at Munich until 1886. The eleven years there were the formative ones of his geographical work.

It is interesting first, perhaps, to see where Ratzel stood during his time at Munich in relationship to German geographical work as a whole. The appearance of Darwin's work and that of his contemporaries had of course deeply affected the development of geography in Germany. Nowhere else is it possible to trace the impact on this subject of what was largely English scholarship, for nowhere else had geography the standing in the universities which it had by then acquired in Germany. Humboldt and Ritter both died in 1859, the year of publication of the *Origin of Species*, and the middle of the century marks the turning away from the concepts of geography which belong so much both to Humboldt's scholarship and, in terms of university work, to Ritter's time at Berlin. The wholeness of the subject, the acceptance of the study of *man* in his environment, the philosophical ingredient inherited from the eighteenth century, and the priming of geography with Ritter's serene, pious teleology, were no longer in keeping with the times. It was the physical geographers who throve, and who now and then, as in the case of Oskar Peschel, began unkindly to expose the weaknesses of the 'Golden Age' of German geography.[17]

How wide the gap was between the two generations of scholars is clear from Merz's comment on Humboldt.[18] He reckons Humboldt as belonging in part to the morphological period of natural science, with his sense of the need to study natural phenomena in their habitat, and to grasp the facts of related and recurrent phenomena whether geological, botanical, or zoological; he sees in this perception not only Humboldt's participation in contemporary science but also the French influence on much of his thought. But he quotes also Humboldt's instinctive side-stepping from the whole theme of evolution: 'The mysterious and unsolved problems of development do not belong to the empirical region of objective observation, to the description of the developed, the actual state of our planet. The description of the universe soberly confined to reality, remains averse to the obscure beginnings of a history of organic life, not from modesty, but from the nature of its object and its limits.' It is possible therefore to see these early nineteenth-century Germans as absorbed by the study of the material world in its physical aspects, but conscious throughout of the great inheritance of philosophical speculation and looking back to the cosmology of the past.

Jean Brunhes wrote later of the difficulty of the task which Ratzel gradually evolved in Munich and later at Leipzig.[19] 'Il est délicat d'observer et d'expliquer les faits naturels: il est bien plus délicat d'observer et d'analyser les faits géographiques humains. Le don d'observation qui est indispensable ne suffit plus. Il est impossible de faire la bonne géographie humaine sans une forte culture historique, économique et philosophique.' Ratzel in his thirties had some of the intellectual equipment needed for this kind of scholarship. The struggle to acquire the missing classical education had provided a grounding in ancient history and archaeology, and with it a deep interest in the existence of simple peoples. Steinmetzler[20] mentions the blank in his learning where medieval history was concerned, with the exception of the famous travellers. But he had an acquisitive mind in which the details and complications of contemporary politics were learned and stored, and a natural taste for philosophical literature.

He could not in any sense be said to have revived Ritter's tradition of academic geography. Certainly he wrote much more gently of Ritter than the geographers of the eighteen-sixties (Part II, 41), and there were reasons for his more liberal attitude towards the early nineteenth-century geographer. As early as 1875, when he began work at Munich, Ratzel had ceased to be a whole-hearted adherent of Darwin's findings, and especially of the theory of natural selection. He remained indeed throughout convinced of the importance of the idea of evolution, and much of his thinking and writing about the application of the idea of organic evolution to human society derived from this absorption of contemporary science. But although the character of his training and interests, and the generation in which he lived, took him away from Ritter's geography, he was not prepared to swallow Darwin's or Herbert Spencer's opinions whole. The same increasingly critical attitude is apparent towards Haeckel, whom earlier he had revered so completely. Haeckel himself disagreed with some of Darwin's findings, although he accepted the more important of them, but Ratzel began to utter reservations about the views of both of them. His comment on Darwin's hypothesis of natural selection in 1870 had been 'der geniale Gedanke der natürlichen Zuchtwahl'. By the end of the century he could criticize it as 'die plumpe Hypothese des Passendsten im Kampf ums Dasein' (Part I, 26, p. 399). The second dictum may have been wholly inappropriate in its wording and indeed presumptuous, but in the mid-twentieth century, and in the years following the Darwinian celebrations, it makes interesting reading, when one bears in mind the controversy that can still kindle over Darwin's work. The whole theme of evolution certainly continued to be fundamental to Ratzel's geographical thinking, but the development of his own concepts of geography took him away from the naturalists and from total acceptance of their findings.

Ratzel had been brought up as a Lutheran, but had quietly discarded belief during his struggle for university education. He wrote of himself as without belief when he went through his military adventures, although he was registered as a Lutheran. This phase lasted, however, just about as long as his unquestioning acceptance of contemporary science, and his whole-hearted return to Protestant Christianity perhaps made him more kindly disposed than his older colleagues towards Ritter's convictions. Finally the philosophical preoccupations which became much more apparent in Ratzel's thinking in middle age made him more sympathetic in his assessment of the early geographers. The effect of Darwin's generation of scientists had been to prune German geography thoroughly of a good many anachronisms, but in the course of it geographical thought had been quite fragmented and the task of reintegration was formidable. Ratzel's concern for human geography was a groping in this direction, in a generation far removed in its outlook from Ritter's. His scholarship was to suffer on account of a later and soured generation from the perversions and humiliations of *Geopolitik* and to attract the reproaches which belong to this pseudo-geographical activity. It is therefore all the more important to see it as accurately as possible in the context of the last quarter of the nineteenth century.

Three aspects of Ratzel's time at Munich are interesting; the lectures that he gave, the books that he wrote, and the friends that he made. Max Eckert, the cartographer who worked under Ratzel later in Leipzig, and who plainly had a great affection and respect for his professor, wrote of the limitations to his lecturing.[21] It was not until his last years at Leipzig, when he knew himself to be one of the outstanding German geographers, who could draw enormous audiences, that Ratzel began to speak with some fluency and feeling for his listeners. Eckert mentions, however, that the monotonous voice was always helped by the careful choice of words, by the solid content of the lectures and by the fact that Ratzel hated repetition of his lectures and always worked to avoid it. Eckert cites the years at Munich as those of the accumulation of Ratzel's vast hoard of learning, sorted and tried out in the form of lectures, and later, if proving worthwhile, developing into books and papers. The list of lecture courses given during the years at Munich and taken from the *Bericht des 'Geographischen Abends'* published later at Leipzig amplifies this point.[22] Ratzel evidently used his American travels freely for his early lectures, but developed a huge range over the ten years. Physical geography, the regional geography of the continents, human and political geography, Antarctic discovery, German travel and travellers of the sixteenth and seventeenth centuries, were all the themes of substantial courses, and formed part of the great store of geographical fact and opinion on which Ratzel was to draw for more specialized work during the first part of his time at Leipzig.

Ratzel will always be remembered as a human geographer, but both by training and conviction, and also probably because of contemporary trends in German geography, he had a grip of and interest in the physical side, and belief in it as an essential foundation to geographical work. This characteristic is plain in his lecture courses (see notes 22 and 27). While he was at Munich one aspect of physical work developed strongly because of his growing affection for Alpine Bavaria, and the amount of time that he spent on field-work in the mountains. The study of the snow levels and of the limestone surfaces in the Alps fascinated him, and he wrote of them steadily (Part II, 66, 112, 113, 114, 115, 118 and 120) and later also at Leipzig. The long comparative geography, *Die Erde und das Leben* (Part I, 23), written towards the end of his life, reiterates this belief in the basic importance of physical geography for all geographical work.

During the eleven years at Munich, Ratzel published five books, and wrote much of the enormous *Völkerkunde* (Part I, 11) which was later to be translated into English under the title of the *History of Mankind* (Part I, 17). Two of these are the descriptive accounts of the United States and Mexico which have already been mentioned (see p. 15). The third was a short comment on those who looked with suspicion or dislike at the growing German inclination to overseas commitments (Part I, 10). The most important was the first volume of *Anthropogeography, or an Introduction to the Application of Geography to History* (Part I, 9).

The second volume was published nine years later with the sub-title, 'The geographical distribution of mankind', and with the hyphen omitted from the significant word: the term had come to stay. A second edition of Volume I came out in 1899. In the two volumes Ratzel expounded his great themes of human geography, those of human societies developing within a frame (*Rahmen*), exploiting a place (*Stelle*), needing space (*Raum*) and finding limits (*Grenzen*).

This treatise on human geography and the immensely long *Völkerkunde*, which belongs to the same period of writing, show many of the problems associated with assessing Ratzel's work. They had, as Professor Broek has pointed out,[23] an immense reputation abroad, and perhaps gave Ratzel a greater standing in the rest of Europe and America than in his own country. *Völkerkunde* was translated into English and Italian, while Ellen Semple's version of the first volume of *Anthropogeography*, though it drew much criticism, was to become famous in the early twentieth century in its formative effect on American geography (see p. 32).

These pieces of writing, expressing Ratzel's search for interpretation as well as the task of description, are full of the phrases and passages which later could suggest that he was a determinist, and Ellen Semple, we know, in her effort to clarify and expound his writing, conveyed that impression.

It fell to the anthropologists, perhaps, and to the American,

Lowie,[24] in particular, to grasp the interest of Ratzel's writing, and to discriminate between the common sense and integrity of his whole approach to an immensely difficult task and the defective vagueness so often apparent in the finished effort. Ratzel's scholarship at Munich was indeed as significant for the anthropologists as for the geographers, since academic anthropology in the field and in literature was then scanty. Ratzel had met the German anthropologist, Bastian, many years before (1868) at Dresden and had studied Herder's much earlier speculative work. But a great deal that he wrote at Munich came from his friendship with and respect for Moritz Wagner, especially the stress laid on the importance of human migration throughout history, and the explanation thereby of the spread and levelling of primitive cultures. This idea he emphasized all through as more reasonable than Bastian's peculiar concept of 'psychic unity', and it came in part from his earnest discussions with Wagner as to the validity of applying to primitive peoples some of the findings of the naturalists about migration in the animal world. Lowie was careful to show, however, that Ratzel never pushed his analogies to the point of underestimating the importance either of human will, or the effect of time. The anthropologists seem to have been much less prone than the geographers to labelling him as a determinist. In *Völkerkunde* Ratzel indeed made a substantial contribution to ethnography—as careful a description as he could formulate, from such literature and field-work as were available, of the contemporary peoples of the earth. The results were inevitably uneven; Lowie mentions critically the absence of sharp demarcations of regional boundaries such as might be expected from a geographer, although Ratzel's successors in this discipline might wonder exactly why such precision should be taken for granted. Ratzel wrote with great attention to Africa, about which the material was more abundant, and gave relatively meagre descriptions of the other continents. That he attempted a range of description which no one could even consider today, and that he had travelled outside Europe for a short while only in America, did not in his generation worry his lecture audiences or his readers. Lowie pointed out that, in 1910, Buschan's further work on the *Völkerkunde* needed five collaborators and that twenty years had brought to anthropology a specialization undreamed of in Ratzel's time.[25] The world-wide scale of study and the reliance on the bits and scraps of field-work and travel literature available were then taken for granted.

Much of his big output of shorter writings emphasized again the preoccupations of these years. Ratzel wrote mainly for the periodical *Ausland* and for *Globus* while he was at Munich, and to a lesser extent for *Meyer's Konversationslexikon* and for the *Österreichische Monatsschrift für den Orient*. He used his American material freely, and, for the Austrian journal, the work that he had done on Chinese emigration. In *Ausland* his growing interest in the Polar regions became apparent, and a new theme, the fate of Africa as a great colonial territory, as well as his preoccupations with the simple societies.

Moritz Wagner stands out as Ratzel's closest friend during his years in Munich and as the scholar with whom he discussed his work most freely and whose advice he valued greatly. Both of them had memories of and interest in American travel. He saw a great deal also of S. Günther, the mathematical geographer who followed him at the Technische Hochschule, W. Götz, who was the geographer at the military academy in Munich, and K. von Heigel, the specialist in Bavarian local history. His steady connection with the historians became much more important at Leipzig, and this supplementing of his own rather patchy scholarship in this field kept the sense of the time factor always apparent in Ratzel's work and possibly also checked the arguments in favour of determinism.

IV

The death of Oskar Peschel at Leipzig, after a very short spell of work, and the move of von Richthofen, the distinguished physical geographer, from that city to take the chair at Berlin in 1886, led to Ratzel's own promotion. Leipzig itself was an attraction: it is well to recall the great vitality and standing of the city in the late nineteenth century. At the time of Ratzel's appointment the long distinction of the university as the largest in Germany was past, for Berlin and Munich were leading in numbers of students. But it had the traditions of an early fifteenth-century foundation and considerable wealth from house property in the city and estates in Saxony for the endowment of teaching and equipment. Moreover, academic life throve, not only in the university but in the wide-ranging and enterprising firms of publishers. Leipzig was at this time the biggest centre of book publishing and selling in the world, and this characteristic had given it an atmosphere far removed from *Kleinstaaterei*, long before the unification of Germany. Karl von Holtei, the actor and writer, had written of Leipzig some years earlier, 'There is only one city in Germany that represents Germany; only a single city where one can forget that one is a Hessian, a Bavarian, a Swabian, a Prussian or a Saxon; only one city where, amid the opulence of the commercial world with which science is so gloriously allied, even the man who possesses nothing but his personality is honoured and esteemed: only one city in which, despite a few narrownesses, all the advantages of a great, I may say a world, metropolis are conspicuous. That city is in my opinion and in my experience Leipzig.'[26] Politically and intellectually, it was pre-eminent in Germany.

Geographical work was already well established, although limited in staff and students, and a chair had been in existence since 1860 for university teaching and for the supervision of the Royal Geographical Institute with its rooms in the *Paulinum* in the Universitätstrasse. The two pieces of writing already quoted, Max Eckert's monograph on Ratzel written in the nineteen-twenties and the *Bericht des 'Geographischen Abends'* published in 1901, leave a reasonably complete account of the running of a German Geographical Institute and the associated university teaching in the late nineteenth century: for Leipzig geography was, perhaps, second only to Berlin's in its reputation. These pieces of work are very much alive, and range from comments on the work and idiosyncrasies of the professor to the regard for Mutter Löchner and her successor, Hermann Kunze, who kept the quarters in the *Paulinum* clean and orderly.

From the point of view of teaching, Ratzel's professorship at Leipzig from 1886 until his sudden death in 1904 was divided into two. For the greater part of the time he was concerned with university geography and had audiences of students which varied in numbers from forty to seventy. In 1900 his lectures were extended to take in the students at the *Handelshochschule* in Leipzig and the audiences became enormous, 300 to 500 at a time. Ratzel evidently stretched himself to meet the demands of this mark of confidence in his teaching, and in the last years of his life his lectures began to make an impression on his students for their style as distinct from their content. The type of lecture course also changed to meet a much more general need. Eckert notes that the fourteen years at Leipzig between 1886 and 1900 saw Ratzel's most specialized work in teaching.[27] Whereas the courses at Munich had roused him to collect a great quantity of knowledge, those at Leipzig enabled him to specialize much more carefully, particularly in the teaching given on Germany and France and the Mediterranean. His teaching became more general again only at the end of his life to meet the syllabuses at the *Handelshochschule*. Throughout he was able to maintain the shorter lecture courses on special topics.

Eckert emphasizes Ratzel's belief in seminar teaching at the Institute and his persevering development of this kind of instruction at Leipzig. The practice of teaching map-reading he had brought from Munich, but his lack of interest and his knowledge of his limitations in cartographical work beyond its elementary stages made him tend to leave the beginners' seminars to his assistants, Fischer, Eckert, Friedrich and, for a while, Hettner. He enjoyed wholeheartedly the graduate seminars and advanced work.[28] In contending with it all he may have been the first, but certainly not the last geographical professor to ache for a larger staff.

One aspect of his labours reads oddly today. Photography as part of geographical equipment was only just beginning to make its appearance, and it was not possible to use it at all freely for formal instruction. A certain amount of skill in freehand drawing was essential for diagrammatic illustration in physical lectures, and Eckert recalls the professor's painstaking efforts to become at all competent in it, as he had no natural talent for drawing.

Two forms of geographical activity he flatly refused to undertake, the organizing of student excursions and frequent attendance at geographical congresses or large gatherings of any sort. He knew quite well the importance of planned field-work for students, and the extent to which periodic gatherings gave his own subject energy and solidarity. He had also given time and thought to a great deal of geographical field-work around Leipzig and in much of highland Bavaria. But he disliked the academic world in large numbers, whether senior or junior, and as far as possible avoided it. (It was possibly this aversion to people in big groups, together with his preoccupation with a load of work, which made him later turn down the suggestion that he should be Rector of the University.) His geographers were left to get what they could out of the geologists' excursions, and a picked few only were asked every year to Schkeuditz, a much prized invitation. Those who saw Ratzel away from his department often enjoyed, too, access to a very happy home,[29] and knew and appreciated the professor as an active and athletic man, a climber and a skier.

Nothing perhaps more showed the attraction of the Leipzig Institute than the marked growth of graduate work there. There were two geographers engaged on research when Ratzel came to Leipzig in 1886, and a steady average of eight to ten annually by the end of the century. Eckert wrote of his professor's kindness to, and anxiety for, his research students, and of his readiness to accept criticism when detailed original work in the field contradicted his own earlier and more generalized studies. In the eighteen-nineties Karl Hassert here prepared his thesis on the Arctic and Eckert his research on the limestone topography of the Jura. During the same years, Ellen Semple came in her early thirties from Vassar College to the Institute and to University lectures. Her study and rendering of Ratzel's work on human geography remain to this day one of the most interesting and perhaps one of the most controversial topics connected with the German geographer.

The interdependence between German and American geographical work has always been marked, although by no means contained within the framework of academic geography. Humboldt made a triumphant return from his South American travels through the States, although he was applauded as much for the professed radicalism of his political opinions as for his scientific field-work and scholar-

A BIOGRAPHICAL MEMOIR

ship. Arnold Guyot, the Swiss, who was to become the first professor of geography at Princeton, was one of Ritter's pupils, and dispensed a little of Ritter's pattern of geographical scholarship to the New World. But much later in the century and indeed up to 1914 two streams of geographical activity, quite separate from each other, were to link Germany and the United States. The one belonged very largely to the physical geographers at Berlin and their work with the American geologists, culminating in the famous exchanges of opinion and teaching between Albrecht Penck and W. M. Davis. They represented a line of work contained wholly within German geography on one side of the Atlantic and mainly within American geology on the other side, and one of the great liabilities to American academic geography of this period was the exclusion from it of substantial physiographical work.

The other stream was conspicuous in Ellen Semple's attention to Ratzel's work, her production of it for the Anglo-Saxon reader and the popularity of her writing for a long period in America. The tale of her experiences in Leipzig in the eighteen-nineties is true to the place and the period. Women were not admitted to University lectures at that date and she listened to Ratzel's formal instruction in a small room opening out of the lecture hall with the door ajar. But she was in the Leipzig seminar long enough and knew Ratzel well enough to develop a real friendship with him and his family,[30] and with it a whole-hearted admiration for his work. It was important, of course, that Ratzel knew and had studied her own country carefully. What she attempted later, to interpret him accurately to the American reader, was perhaps impossible. In the Introduction to the *Influences of Geographic Environment*[31] she explained the difficulty that any informed German would have had in following completely what was rightly assessed as a landmark in scientific writing, but also as a confused and muddled piece of work, a courageous groping rather than an achievement. Her rendering after the passage of time, and in another language, gave an impression which hardly did justice to Ratzel's total and developing approach to environmental influences, just as interpretations of a damaging kind of his political geography and biogeography could later give the impression that the geopoliticians were in line with the German geographers.

What is written here should not be taken as belittling Ellen Semple's standing as a scholar. This side of the German–American connection in geographical work is not like that between the geologists and the physiographers. In human geography the traffic was one way only: what Ellen Semple learned and wrote of Ratzel's work made a profound impression on American academic geography, even if the Americans, as their own work developed, were to find the apparent determinism of Ratzel's writing and hers an endless topic of controversy. But there is no account in German of Ellen Semple's coming to Leipzig, nor of any American contribution to German geographical thought at the time on the human side. To contemporary German geographers she was another foreign graduate studying at the Leipzig Institute. Yet if Ellen Semple's rendering of a problematical piece of writing seems (forty years later) to mislead, who has done any better? What would many Anglo-Saxon readers know at all of Ratzel's *Anthropogeography* if it were not for her interpretation of it? And what had she at the time in the way of contemporary and relevant American scholarship to guide her? It was as a historian and economist that she went to Leipzig in the eighteen-nineties, because the Geographical Institute there was already famous, and because in the American universities there appeared to be no possibility whatever of studying the theme of environment which interested her so deeply. Moreover her own comment in the preface already quoted shows her awareness of the pitfalls of determinism, and of the doubtful validity of the organic theory of state and society.

Another feature of Ratzel's work in Leipzig was the founding and organizing of the 'Geographischer Abend', the name by which the Leipzig Geographical Society came to be known. In its short, written constitution[32] it is described as 'a free Society of Leipzig gentlemen who are engaged in the scientific study of geography', and in its membership it went far beyond the University to interest the businessmen of Leipzig and especially the publishers. Ratzel made it clear, as he had done in a stricter and more limited fashion in the Institute, that students of kindred subjects (in this case both within and without the University) were welcome as long as they had a genuine interest in geography. The names of the members of the 'Geographischer Abend' over the years from other parts of the continent and from overseas show again the international reputation which Ratzel gradually built up for Leipzig geography.[33]

In the list of local members, the link with the Leipzig

publishing houses is also worth remembering, especially with Brockhaus and with Velhagen and Klasing. In nineteenth-century Germany geography as a whole, one great source of strength to the faculty and one reason for the articulateness of the geographers was the close link with certain publishing houses. The outstanding case is that of Justus Perthes at Gotha, and of the early and constant attachment of that great house to geographical scholarship, but the association is again well brought out during the years of Ratzel's work at Leipzig.

The other big element among Leipzig members was that of the teachers, and for his generation of University professors and for so busy a man, Ratzel was perhaps precociously concerned with the quality of teaching in his subject in primary and secondary schools. The encouragement to it was the high standard of teaching in Saxony as a whole, but Ratzel's was not wholly an academic preoccupation. The characteristic which irritated some of his colleagues, an outspoken affection for Germany, came out in the anxiety that German children of all classes and educational standards should grow to love their country because they had studied it carefully. Once again it is necessary to see this side of geographical work objectively, and to realize how instinctive is the suspicion of the regimentation of the German schools which became a fact a generation later. Yet in itself this close connection in Saxony between the University geographers and the schools had everything to commend it and, in a small but quite definite way, the effort to get across to the school population is reflected in Ratzel's writing (Part I, 19 and Part II, 226, 270).

A BIOGRAPHICAL MEMOIR

Ratzel produced an enormous amount of written work between his appointment to the chair at Leipzig and his death in 1904. One aspect of it was the accumulation rather than the consummation of his publications on human geography. The most important of them were the second volume of *Anthropogeography* and the massive comparative geography, the *Earth and Life*, which has sometimes been compared with Réclus' general geography, *La Terre*. It has been described as Ratzel's most mature work; and certainly there is the compound in it of the many elements of scholarship which predominated during the various phases of his life. It reiterates, as has been noted (p. 22), his firm adherence to a physical basis for geographical work, and shows the abiding importance for Ratzel of the fundamental ideas of the mid-nineteenth-century natural scientists. It shows his wide study of contemporary philosophy and is something of a forecast also of his final concentration on biogeography. Certainly in the last ten years of his life, Ratzel's interest in philosophy grew steadily and is apparent in much of his writing. Whereas at Munich he had still been preoccupied with the findings of the natural scientists, at Leipzig his concepts of human geography in general, and of political geography in particular, were complicated by his attempt to correlate philosophical with geographical thinking. Sauer summarized this development in his comment on Ratzel's philosophical belief in two kinds of scientific learning, that of actuality and that of the abstract.[34] Ratzel's tendency at Leipzig was to urge the relationship of geography with the latter, and Sauer saw this preoccupation as producing something like a stalemate in much of

Ratzel's work. Comte's writing, and that of Spencer, Fechner and Kopp and the historian Buckle, were read and pondered over, although in no case wholly accepted, and this characteristic of Ratzel's work in its later stages perhaps sets his approach to geography once more in line with that of Ritter. The impression grows with Ratzel's published and quite emphatic reassertion of his religious beliefs in the *Deutsche Monatsschrift* in 1901 (Part II, 300) and in *Glauben und Wissen* in 1903 (Part II, 318).

For both geographers and anthropologists, one shorter writing, the article on 'Nationalities and Races', published in the *Türmer-Jahrbuch* (Part II, 326) is significant, for it contains an explicit warning against the racial theories of Gobineau and Houston Chamberlain. It was already plain in Ratzel's lifetime where perversions of the rapidly growing interest in race distinctions might lead, and there is no doubt either of the rejection by his generation of geographers of these dangerous possibilities.

Perhaps the greatest single emphasis in these years was on political geography (Part I, 15, 18, 22, and Part II, 218, 254, 284). The numbers cited are those of longer works; in addition, the bibliography shows the enormous range of comment on the political problems of the day and the contemporary emphasis on the scramble for Africa, the developments in South Africa and the growing preoccupations with competitive naval strengths.

Seen as a whole, Ratzel's writings on political geography show the two main approaches to this kind of geographical work, and it is difficult to think of another case in which *both* appear so abundantly and seem to have been mastered by one man. Ratzel's knowledge of contemporary political events and their relevance for the geographer was almost always expressed in short articles like those in the *Grenzboten* and the supplements to the big dailies. Only in one major work, the second on the United States (Part I, 15), does it appear at length in regional form, and only once in an indirect way in topical form, with his study of Emin Pasha (Part I, 12), whose activities evidently had a fascination for him. Some of these articles show Ratzel at his noisiest (201, 242, 262) and the unabashed aggressiveness of his style shows exactly what might provoke the assumption that he was the propagator of an overt geographical imperialism. One or two instances of writing and of anecdote should however make one think twice. At Munich (1878) Ratzel wrote an article for *Nord und Süd* (Part II, 40) in which he discusses the character of the German minority settlements in Europe, beyond the German frontiers. One passage in it is so interesting as to be worth quoting (always with the hesitancy with which geographers now risk quoting Ratzel)—'And I frankly think that we Germans ought to be glad that independently active and productive groups of our people are preserved for us in Switzerland, in Austria and in the Russian Baltic provinces. These people, who are detached politically, but who have remained part of us in mind in quite other circumstances, think and feel in some ways quite differently from ourselves. While it is doubtful whether their reunion with us would make us stronger, it is certain that these same groups would not enrich (by return) our German cultural existence: it would just become more uniform.' In evaluating the German minority populations this is a long way from the shrill cries during the Third Reich for their compulsory return to Germany.

Max Eckert speaks of another instance while Ratzel was at Leipzig which has its significance in the mid-twentieth century. Political feeling ran high in Germany during and after the Boer Wars and much of it, official and unofficial, was anti-British. When the Orange Free State and the Transvaal were annexed by Great Britain, many of the senior members of Leipzig University signed a public protest against British policy and action. These included Ratzel's personal friends, for example K. Lamprecht, the historian, with whom he was accustomed to discuss closely much of the purpose and content of his work (see p. 43). When asked to add his signature Ratzel declined, saying that, on grounds of common sense, he could not see that the South African Dutch had it in them to make a success of these territories, and that there was therefore no point in making the protest. His refusal roused a good deal of incredulous anger as it might in some quarters today.

His abundant writing on contemporary politics shows at times that pugnacity in style and that tendency to speculation which were to lend themselves later very easily to perversion. It also shows an enormous amount of factual knowledge and a sort of rough and independent common sense in judgment which takes him right away from the mystical geopoliticians of the twentieth century. Dr Parrella's thesis on *Lebensraum and Manifest Destiny*[35] makes interesting reading at this point. He has no difficulty in saddling Ratzel and even Ritter with hatching, if unconsciously, the tendentious thinking which later would develop into geopolitics. This attempt to trace damaging political ideas throughout nineteenth- and early twentieth-century German academic geography is not easy to accept. But Parrella works out (always assuming that he is not also overharsh with his fellow countrymen) something like a parallel connection between the popular version in America of the idea of the expanding organism and the contemporary political theme of Manifest Destiny. Some of the excerpts quoted in his thesis are startling enough.[36] We are pondering now, a century after Darwin's publication of the *Origin of Species*, over the idea of a long cycle in which socio-political theory (in part that of the French Encyclopedists) stimulated biological theory (that of evolution) which in its turn created another socio-political pattern (that of geopolitics), to be modified again by more biological findings (the work of the ecologists). It can be argued that this to-and-fro between scientific and political thought goes back well into the eighteenth century and it is thus possible to see the outbursts of perversion involved in proportion and with more detachment. But geographers might wonder whether Ratzel's thought and work in the setting of Germany at the turn of the century might not easily be misconstrued into a hideous pattern of ideas; while, in contrast, the same instincts in the New World, far removed from the tradition of the Balance of Power in Europe, and with nothing but the inhibitions of the American people to check expansive expression and operation, could rise to a climax and subside without calamity. Cord-Meyer, in his long study of *Mitteleuropa*,[37] certainly makes it clear that he considers no German geographer before 1914 to have been deliberately subjected to pressure or tempted to use his scholarship for political ends, unless it were at Graz in Styria. What he writes also does something to dispel the illusion that academic geography in Germany really had the popularity and standing which are sometimes claimed for it, and suggests that it was the dangers and hardships of the First War which made most Germans aware for the first time of the practical importance of position and resources. Ratzel certainly made no secret of his dislike of party politics. He was deeply interested in foreign affairs and the relevance of geographical comment to them, but had neither time nor the wish to go beyond this kind of correlation.

For the geographers, Ratzel remains the founder of political geographical studies. He had the range of relevant geographical learning which sets him apart from the earlier speculative writers and indeed from the nineteenth-century philosophers. Of Spencer he was to complain 'Von dem Boden als naturgegebenes Kontinuum hat Spencer keine Ahnung' (Part II, 232), and the same comment applied to Comte's reflections. For the practical purposes of the political geographer, the two thinkers, much as they were revered and cited, simply did not bring the conviction which belonged to the long years of accumulating, sifting and analysing the facts of position and resources.

The big treatise (Part I, 18) that belongs to these years forms a different approach to the subject: the attempt, based in part on the accumulation of descriptive work and in part on philosophical preoccupations, to give this relationship between the earth and the state some theoretical foundations. The published results, like those in human geography, tend to baffle rather than to inspire and seem tedious enough today. It is only after looking at the meagre achievements in political geography since Ratzel's time that the geographer begins to feel more respectful towards his efforts. Very few men have since then come near to his knowledge of contemporary politics. Twentieth-century political geography tends inevitably to attract compilations, and Bowman alone, perhaps, had something of the same grasp of the whole field of political issues. Fewer still have been able to attempt to discern the critical topics and ideas which might in time give this application of geographical fact and technique validity and reliability. The intellectual energy which in a massive rather than a disciplined form set political geography on its troubled way has really had no counterpart in the last half-century, and the notoriety of geopolitics has scared a good many geographers away from the subject. Indeed, the time is now past when a single scholar could ever hope to master the range of learning either in descriptive or argumentative form.

One other short piece of writing towards the end of his life, 'Der Lebensraum' (Part II, 295), has given Ratzel much of his later unenviable reputation as the geographical sponsor of German imperialism. No less a person than Goethe has been charged with the first use of this expression;[38] and interesting as are Goethe's occasional plunges into geographical reflections, and his friendship with, and interest in Humboldt, there is nothing sinister about his ruminations. It is well to read Ratzel's 'Der Lebensraum', and perhaps recognize in it something of a reversion to earlier ideas. This is not at all an admission of determinism, but a pondering as to whether the development of biogeography with its emphasis on plant and animal distributions, and the constant movement of, and adjustment between different species, was not the critical link between physical and human geography. There is frequent reiteration and an elaboration of themes that he and Moritz Wagner had so often discussed in Munich: the need to study not only the movement of plant and animal and human life, but also the power of all these things to settle and establish themselves in new environments. One or two passages in 'Der Lebensraum' show so exactly what might happen when they were interpreted by a generation both soured and emotional, that they are worth quoting.[39] But to his contemporaries, and certainly for a more detached generation of geographers, it is probably right to assess this piece of writing as a struggle towards thinking out the scope and content of biogeography. Brunhes[40] described Ratzel as not philosopher, nor historian, nor ethnographer, nor economist, but rather a naturalist, discerning and reckoning the effects on society of physical feature, altitude, topography, climate and vegetation. Lamprecht, the historian, saw Ratzel's significance and success as a human geographer in his recurrent search for the links between physical phenomena and human activity through biological and biogeographical studies. This is an assessment also which French human geographers of our own time[41] have been quicker to value than those of any other people. As in political geography, the difficulties besetting biogeographical work of which we are conscious today suggest that Ratzel had grasped the general idea of another approach to geography, vital to its integration, but

A BIOGRAPHICAL MEMOIR

that he got no farther than roughing out the lines of study. No geographer since has found it an easy form of learning.

Steinmetzler[42] is careful to point out that Ratzel's teaching and writing during the years at Leipzig were those of his own maturity, and by that time owed much less to contemporary scholarship. Nevertheless his friendships at Leipzig are interesting, and especially his close association with Karl Lamprecht, with Wilhelm Wundt, the philosopher, and with Wilhelm Ostwald, the chemist. Steinmetzler also mentions Ratzel's respect and liking in these years for Rudolf Kittel, the professor of theology. Some of Lamprecht's concepts of history,[43] the effort to correlate it with contemporary scientific method and to emphasize phases of civilization in terms of collective psychical conditions, suggest the problems which also dogged Ratzel. Lamprecht's inclination to make less of individual achievement, to stress trends in economic history, and to attempt an appraisal of environment, may reflect the exchanges of thought and opinion between him and the three other scholars.

The four men were accustomed for some years to meet on Friday evenings, as Brunhes says, 'pour causer ensemble de ces problèmes philosophiques fondamentaux de la nature et de l'homme auxquels nous nous heurtons bon gré mal gré, aux confins de toutes les sciences'.[44] What remains interesting for a later generation of geographers is the emphasis *not* on the definition and scope of any discipline, but on the confident assumption throughout of the enormous amount of common ground for discussion.

Ratzel died suddenly from a stroke in the late summer of 1904, a few weeks before his sixtieth birthday, while he and his family were at their summer home in Ammerland in Bavaria. It is known that he had in mind a development of the biogeographical themes which he anticipated in *Die Erde und das Leben* and sketched out in 'Der Lebensraum' and, if he had lived, this might have been the final dominating idea of his work. Whether further years of activity would have given his thinking and writing any greater discipline is doubtful, and it was his hard fate to be remembered by geographers at large through the generation that misused his work rather than by the usefulness of the scholars whom he trained at Leipzig.[45] He had throughout, as Lowie insisted, a capacity for conceiving comprehensive ideas coupled with a comparative deficiency in the formulation of definite problems,[46] and it seems unlikely that a little more time would have altered the much criticized and congenital shapelessness of a good deal of his work. Yet if he was one to follow, rather than to imitate, he stands out amongst German geographers for the range and vigour and courage of his scholarship. Herbert Spencer's sententious dictum on the scientists who preceded Darwin[47] applies quite truthfully to Ratzel and his laborious efforts to assess environment: 'These speculations, crude as they may be considered, show much sagacity in their respective authors and have done good service. Without embodying the truth in a definite shape they contain certain adumbrations of it. Not directly, but by successive approximations do mankind reach correct conclusions; and those who first think in the right direction, loose as may be their reasonings, and wide of the mark as their inferences may be, yield indispensable aid by framing provisional conceptions, and giving a bent to inquiry.'

A BIOGRAPHICAL MEMOIR

SUGGESTIONS FOR FURTHER READING

Books

Geography in the Twentieth Century, ed. Griffith Taylor (London, Methuen, 1951).

Darwin and Modern Science, ed. A. C. Gerrard (Cambridge, at the University Press, 1909).

G. R. Crone, *Modern Geographers* (London, the Royal Geographical Society, 1951).

Articles

Geographical Journal, vol. XXIV (1904). Obituary notice of Friedrich Ratzel.

O. Marinelli, 'Frederigo Ratzel e la sua opera geografica', *Rivista geografica Italiana*, vol. XII (1905).

NOTES

1. J. Gottmann, *La politique des états et leur géographie* (Paris, Librairie Armand Colin, 1952), ch. II; Y. Goblet, *Political Geography and the World Map* (London, George Philip, 1959), pp. 8–12.
2. J. O. M. Broek, 'Friedrich Ratzel in Retrospect', abstract in *Annals of the Association of American Geographers*, vol. XLIV (1954), p. 207. The full version of the paper was privately circulated.
3. The periodical *Grenzboten* was founded in 1841 by Ignaz Karanda in Brussels and represented originally the views of a German–Flemish Liberal group. Very early in its existence it came into the hands of the Leipzig publishing firm of Grunow. Between 1861 and 1870, Gustav Freytag, the Silesian Liberal, was the editor, and the *Grenzboten* spoke for the German 'Bürgertum'. In the eighteen-nineties its outlook was markedly conservative, but there was no party commitment associated with it.
4. A. Hettner, Reviews of *Glücksinseln und Träume* and *Kleine Schriften* in the *Geographische Zeitschrift*, vol. XIII (1907).
5. H. Schrepfer, 'Was heißt Lebensraum?', *Geographische Zeitschrift*, vol. XLVIII (1942).
6. H. Beck, *Alexander von Humboldt* (2 vols., Wiesbaden, Franz Steiner, 1959); R. Bitterling, *Alexander von Humboldt*, (Munich and Berlin, Deutscher Kunstverlag, 1959); *Alexander von Humboldt. Studien zu seiner Universalengeisteshaltung*, ed. J. H. Schultze (Berlin, Walter de Gruyter, 1959); Also, H. de Terra, *Alexander von Humboldt* (New York, Knopf, 1955); *Die Erde* (90th annual issue), Part II, devoted to papers on K. Ritter; and W. L. Gage, *The Life of Karl Ritter* (Edinburgh, Blackwood, 1867).
7. 'Lebenslauf von Friedrich Ratzel' (1844–1869), *Kleine Schriften*, vol. 1.

A BIOGRAPHICAL MEMOIR

8 Quoted by K. Hassert, 'Friedrich Ratzel, Sein Leben und Wirken', *Geographische Zeitschrift*, vol. XI (1905).
9 *Ibid.*
10 Gage, *op. cit.* pp. 139–40.
11 J. Brunhes, 'F. Ratzel, 1844–1904', *La Géographie*, vol. X (1904), Paris, Massons et Cie.
12 C. Darwin, *More Letters* (London, John Murray, 1903), vol. I, p. 391.
13 The following books and newspapers are cited in *Die Chinesischen Auswanderungen*: Sir J. Bowring, *The Kingdom and People of Siam*; Burney, *Burma and its People* (1860); Cooper, *Travels of a Pioneer of Commerce* (1872); J. Crawford, *Journal of an Embassy to the Courts of Siam* (1828); James Elgin (Earl of Elgin), *Letters and Journals* (1875); B. Fortune, *Residence among the Chinese* (1857); W. H. Medhurst, *The Foreigner in Far Cathay* (1872); W. W. Smyth, *The Gold Fields of Victoria* (1869); Williamson, *Travels in N. China* (1870); *Journal of the Indian Archipelago*; *London and China Telegraph*.
14 See O. Howarth, Presidential address given to Section E of the British Association, Edinburgh, 1951.
15 A. von Humboldt, *Essai politique sur le royaume de la Nouvelle Espagne* (Paris, 1811); *Essai politique sur l'île de Cuba* (Paris, 1826).
16 Gage, *op. cit.* p. 191, and J. Kramer, 'A note on Karl Ritter', *Geographical Review*, vol. XLIX (New York, 1959).
17 It should not be forgotten, however, that Ratzel's seniors, two of whom were at Leipzig, produced substantial works on human geography which are listed here:

F. von Richthofen, *China: Ergebnisse eigener Reise*, 5 vols. (three published posthumously) (Berlin, Dietrich Heimer, 1877–1911). The emphasis is on physical geography, but there are long sections on human geography.

O. Peschel, *Völkerkunde*, Leipzig, 1877.

A. Kirchoff, *Vorlesungen für die Geographie, Mensch und Erde*.

18 J. T. Merz, *A History of European Thought in the Nineteenth Century* (Edinburgh, 1903), vol. II, p. 277. The quotation is from the section on organic life in vol. I of *Kosmos*.
19 Brunhes, *op. cit.* p. 106.
20 J. Steinmetzler, *Die Anthropogeographie Friedrich Ratzels und ihre ideengeschichtlichen Wurzeln* (Bonn, 1956), p. 70.
21 M. Eckert, 'Friedrich Ratzel', *Verfasser*, Breslau, 1927.
22 *Bericht des Geographischen Abends (Vereinigung von Leipziger Geographen), zugleich Festschrift zur Feier der 25-jährigen Universitätslehrtätigkeit als Professor des Begründers des Geographischen Abends, des Geh. Hofrats Prof. Dr. Friedrich Ratzel* (Leipzig, 1901). The list of lectures here and in Note 27 is not complete, since many were repeated or given again in a slightly varied form. It serves to show the range of topics covered:

1875–76 America
Comparative geography of high mountain regions
1876–77 General geography of Europe
Commercial geography
The colonies of the European powers
Physical geography
1877–78 Human geography
1878–79 The polar regions
America
Australia
Asia and Africa
1880–81 Great journeys of discovery
1881–82 Australia and Polynesia
1883–84 The peoples and countries of Asia
1884–85 The American peoples
Anthropogeography

A BIOGRAPHICAL MEMOIR

            The geography of and journeys of discovery in Antarctica
  1885–86 The geography and ethnography of Africa
            Principles of political geography
            German travellers and travel writers of the sixteenth and seventeenth centuries.

23   J. Broek, *op. cit.*
24   R. H. Lowie, *The History of Ethnological Theory*, New York, 1937.
25   *Ibid.* p. 122.
26   See the very informative outline account of Leipzig in the thirteenth edition of the *Encyclopædia Britannica* in which Hassert, the geographer, is listed as one of the authorities on the history of the growth of the city.
27   Lecture courses given at Leipzig 1886–1902. (The Report of the Leipzig Geographical Society was published in 1901, so that the courses listed up to 1900 show the more specialized work, and only the beginning of the resumption of more general work.)

  1886–7 Anthropogeography
           Snow, firn and glaciers
  1887–8 Selected topics of Alpine geography with instruction from observation
           The colonization of the African lands
           Introduction to political geography
           The political geography of Europe
  1888–89 Anthropogeography with demonstrations in the Museum für Völkerkunde
           Germany and her neighbours
           Non-European states and their colonies
           The Alps
  1890–91 General political geography with special reference to European states and their colonies
           Africa
           Germany and the territory of the Germans

1891–92 General geography, including the principles of biogeography
The United States
1892–93 The Geography of Germany and France
Survey of the natural and demographic conditions of Africa
1893–94 The Mediterranean lands and peoples
Physical geography
Survey of the most important European states
1894–95 The morphology of the earth
Principles of political ethnography
1896–97 Climatology and hydrography
Survey of the literature of travel
Natural and historical landscape
1897–98 The influence of environment on history
England's world power and policy
Landscapes and cities of Central Europe
1898–99 Biogeography
Germany and Central Europe (photographic illustrations)
The most important transport routes
1899–1900 Introduction to morphology and hydrology (photographic illustrations)
Oceanography and climatology
The chief commercial centres outside Europe, and German political and trading relations with them
The principles of landscape and nature study (photographic illustrations)
1900–1901 Germany and German Central Europe
General geography: the parts of the earth, soil types
The lands and peoples of Europe
1901–1902 Geography of fens and rivers
Transport geography
The science of the assessment of peoples

28 Subjects for seminar discussion at the Leipzig Institute (1896–1901):
    Selected topics of physical geography
    Themes in the geography of Central Europe
    Selected passages from the writings of Reinhold Forster, Alexander von Humboldt and Karl Ritter
    Selected topics of anthropogeography
    Practical work for advanced students in geomorphology
    Discussions on the physical geography of Germany
    Exercises and discussions on oceanographical charts
29 Ratzel married Marie Wingens, and there were two daughters of the marriage. Very little information about his family life is available.
30 Ellen Semple's friendship with the widow and daughters continued after the First War, and enabled her to help them during the inflation of the nineteen-twenties in Germany.
31 E. Semple, *The Influences of Geographic Environment*, (London, Constable, 1935 edition).
32 See the *Bericht des Geographischen Abends*.
33 Foreign members of the *Geographischer Abend* to 1901: Danoff (Bulgaria); Darbishire (U.K. (Oxford)); De Martonne (France (Rennes)); Douffet (India (Calcutta)); Gukassian (Caucasia); Iwanowsky (Russia (Moscow Historical Museum)); Kalpaktschieff (Bulgaria); Krug-Genthe (U.S.A. (New York)); Magnus (Norway (Bergen)); Mehedinti (Rumania (Bucharest)); Mondaini (Italy); Monroe (U.S.A. (Westfield, Mass.)); Pflug (Russia (Riga)); Popovic (Serbia (Belgrade)); Sapundieff (Bulgaria); Semple, Ellen (U.S.A. (Louisville)); Smiljanic (Serbia (Belgrade)); Walser (Switzerland (Bern)).
34 C. Sauer, *Encyclopaedia of the Social Sciences*, vol. XIII, New York, The Macmillan Co., 1949.
35 F. Parrella, *Lebensraum and Manifest Destiny* (Georgetown, 1957).

36 For example, Francis Lieber in 1838 considered the state as 'an organism; that is, a moral unity which originated in and ever rested upon man's social instinct'. The *Democratic Review*, 1847: 'The Mexican race now see in the fall of the aborigines of the North their own destiny. They must amalgamate or be lost in the superior vigour of the Anglo-Saxon race, or they must utterly perish.' Senator Benton in 1861: 'The White race went for land and they will continue to go for it where they can get it. Europe, Asia and America have been settled by them in this way.... The principle is founded in their nature and will continue to be obeyed.' Most interesting of all, perhaps, is the quotation from Charles Darwin: 'There is apparently much truth in the belief that the wonderful progress of the United States, as well as the character of the people, are the results of natural selection: the more energetic, restless and courageous men having emigrated during the last ten or twelve generations to that great country and having there succeeded best.' *The Descent of Man*, vol. 1, p. 179 (London, John Murray, 1871).

37 H. Cord-Meyer, *Mitteleuropa in German Thought and Action: 1815–1944* (The Hague, Martinus Nijhoff, 1955), pp. 110, 140, 244.

38 H. Schmitthenner, *Lebensräume im Kampf der Kulturen*, (Heidelberg, 1957).

39 'Between the movement of life which is unceasing, and the extent of the earth which is unalterable, there is a struggle (*Widerspruch*). And from this struggle the war for space was born' (p. 153). 'A people will not stay for generations in the same territory: because it increases, it must expand' (p. 172).

40 Brunhes, *op. cit.* p. 104.

41 M. Sorre, *Les Fondements de la Géographie Humaine*, (Paris, Librairie Armand Colin, 1947); *La Migration des Peuples* (Paris, J. Flammarion, 1955); *Rencontres de la*

Géographie et de la Sociologie (Paris, Librairie Marcel Rivière et Cie, 1957).

42 Steinmetzler, *op. cit.* p. 143.
43 K. Lamprecht, *What is History?* (trans. Andrews) (New York, Macmillan, 1906).
44 Brunhes, *op. cit.* p. 107.
45 Amongst the geographers should be reckoned: Eckert, later to work in Kiel; Viktor Hantzsch, the compiler of the bibliography; Hassert, to work in Cologne; Oberhümmer, in Vienna; Ellen Semple at Clark University, U.S.A.; Werle in Leipzig. Hans Helmolt, the historian, was also Ratzel's student at Leipzig.
46 Lowie, *op. cit.* p. 216.
47 H. Spencer, *Principles of Biology* (London, Williams and Norgate, 1898), vol. I, p. 496.

PART II

A BIBLIOGRAPHY OF RATZEL'S WORK

Viktor Hantzsch's bibliography of Ratzel's written work is printed in the second volume of the *Kleine Schriften*, a collection of some of Ratzel's shorter writings published by his friend, the historian Hans Helmolt, two years after the geographer's death. The bibliography is in two parts, a shorter list of titles of complete books, and a much longer one of contributions to journals and newspapers, both arranged in chronological order. The same arrangement has been preserved in the following English version. Hantzsch's list runs to 543 titles in the second part, and a good many of these have been omitted in this work: most of the deleted items are obituary notices, a form of interest in his fellow men for which Ratzel apparently had a curious enthusiasm and which was known as one of his eccentricities. Entry 199 in Part II, 'The deterioration of necrology', expresses his disgust at the decline in the standard of this particular type of writing.

The list is not quite as long as it looks for various reasons. Some second editions are listed. The articles in the *Kleine Schriften* and in another anthology, *Glücksinseln und Traüme* (Part I, 26 and Part II, 336), are collections of previous writings, and those items in Part II of the bibliography which appear in the *Kleine Schriften* have been marked with an asterisk. In Part I, the *History of Mankind*

(17) is the English translation of *Völkerkunde* (11); and between 1898 and 1900 the list of articles on Corsica contains items in translation which are almost identical (Part II, 272, 273, 289).

The figures in square brackets that follow the main numbers for each item are the numbers in Hantzsch's original list (the first five entries in Part I are identical in numbering), and the German titles, in round brackets, follow the English translation. Hantzsch's bibliography is not easily accessible to readers, as copies of the *Kleine Schriften* are rare in this country. But it remains the authoritative list, and it is necessary for a shorter English version to maintain the connection with the original.

The paging of books and of almost all articles is also printed, as it gives on the one hand, especially in Part I, an impression of the enormous bulk of Ratzel's written work, and on the other an idea of the extraordinary diffusion of scraps of activity over a huge range of topics and publications. It has seemed better to give the short vocabulary of German terms common in printing and publishing that follows rather than to attempt translations of the particulars of publication.

Abhandlung	paper
Auflage	edition
Aufsatz	essay
Beilage	supplement
Bericht	report
Bild	sketch (literally 'picture')
Blätter	papers (literally 'leaves')
Folge	series
Halbmonatsheft	fortnightly review, number

herausgeben	to edit, publish
Jahrbuch	year-book
Jahresbericht	annual report
Jahrgang	annual issue
Monatsheft	monthly number
Monatsschrift	monthly magazine
Sitzungsbericht	report of a committee
Verhandlung	transaction
Veröffentlichung	publication
Verzeichnis	bibliography
Vierteljahrsheft	quarterly number
Zusatz	appendix

I. BOOKS

1 *The Organic World, its Present State and its Development: a Popular History of Creation.* (*Sein und Werden der organischen Welt. Eine populäre Schöpfungsgeschichte.*) Leipzig: Gebhardt and Reisland, 1869. Pp. 514.

2 *Travels of a Naturalist.* (*Wandertage eines Naturforschers.*)
  Vol. I: *Zoological Letters from the Mediterranean: Letters from South Italy.* (*Zoologische Briefe vom Mittelmeer. Briefe aus Süditalien.*) Pp. 333.
  Vol. II: *Sketches from Transyvania and from the Alps.* (*Schilderungen aus Siebenbürgen und den Alpen.*) Leipzig: F. A. Brockhaus, 1873–4. Pp. 282.

3 *The Pre-History of the European peoples.* (*Die Vorgeschichte des europäischen Menschen.*) Munich: R. Oldenbourg, 1874. Pp. 300.

4 *Sketches of the Cities and Culture of North America.* (*Städte und Kulturbilder aus Nordamerika.*) Leipzig: F. A. Brockhaus, 1876. Part I, pp. 258; part II, pp. 265.

5 *The Chinese Emigration. A Contribution to Cultural and*

Commercial Geography. (*Die chinesische Auswanderung. Ein Beitrag zur Kultur- und Handelsgeographie.*) Breslau: J. U. Kerns Verlag, 1876. Pp. 272.

6 [7] *The United States of America*. (*Die Vereinigten Staaten von Nordamerika*.)

Vol. I: *Physical Geography and Natural Characteristics*. (*Physikalische Geographie und Naturcharakter*.) Pp. 667.

Vol. II: *The Cultural Geography of the United States of North America with Special Consideration of Economic Conditions*. (*Kulturgeographie der Vereinigten Staaten von Nordamerika unter besonderer Berücksichtigung der wirtschaftlichen Verhältnisse*.) Munich: R. Oldenbourg, 1878, 1880. Pp. 762.

7 [8] *From Mexico: Travel Sketches of 1874 and 1875*. (*Aus Mexiko. Reiseskizzen aus den Jahren 1874 und 1875*.) Breslau: J. U. Kerns Verlag, 1878. Pp. 426.

8 [10] *The Earth: Twenty-Four Popular Lectures on General Geography. A Geographical Text-Book*. (*Die Erde, in 24 gemeinverständlichen Vorträgen über allgemeine Erdkunde. Ein geographisches Lesebuch*.) Stuttgart: J. Engelhorn, 1881. Pp. 440.

9 [11] *Anthropogeography, or an Introduction to the Application of Geography to History*. (*Anthropo-Geographie oder Grundzüge der Anwendung der Erdkunde auf die Geschichte*.) Stuttgart: J. Engelhorn, 1882. Pp. 506.

10 [12] *Against the Fault-Finders of the Reich. A Word from Electors on the Colonial Question*. (*Wider die Reichsnörgler. Ein Wort zur Kolonialfrage aus Wählerkreisen*.) Munich: R. Oldenbourg, 1884. Pp. 32.

11 [13] *Ethnology*. (*Völkerkunde*.)

Vol. I: *The Primitive Peoples of Africa*. (*Die Naturvölker Afrikas*.) Pp. 660.

Vol. II: *The Primitive Peoples of Oceania, the Americas and Asias*. (*Die Naturvölker Ozeaniens, Amerikas und Asiens*.) Pp. 815.

Vol. III: *The Civilized Peoples of the Old and New Worlds*.

BIBLIOGRAPHY

(*Die Kulturvölker der Alten und Neuen Welt.*) Leipzig: Bibliographisches Institut, 1885, 1886, 1888. Pp. 779.

12 [14] *Emin Pasha. A Collection from the Travel Correspondence and Reports of Dr Emin Pasha from the Former Equatorial Provinces of Egypt and the Lands Bordering on them.* (*Emin Pascha. Eine Sammlung von Reisebriefen und Berichten Dr. Emin Paschas aus den ehemals ägyptischen Äquatorialprovinzen und deren Grenzländern.*) Leipzig: F. A. Brockhaus, 1888. Pp. 550.

13 [15] *Snow-Cover, with Special Reference to the Mountains of Germany.* (*Die Schneedecke, besonders in deutschen Gebirgen.*) Stuttgart: J. Engelhorn, 1889. Pp. 170. A volume in the series of research studies on German 'physical' and human geography produced under the auspices of the Central Commission for scientific information on Germany and edited by A. Kirchoff. (Vol. IV, Part 3.)

14 [18] *Anthropogeography. Part II. The Geographical distribution of Mankind.* (*Anthropogeographie. 2. Teil. Die geographische Verbreitung der Menschen.*) Stuttgart: J. Engelhorn, 1891. Pp. 781.

15 [19] *The United States of America. Vol. II. The Political Geography of the United States with Special Consideration of Natural and Economic Conditions.* (*Die Vereinigten Staaten von Amerika. Bd. II. Politische Geographie der Vereinigten Staaten von Amerika unter besonderer Berücksichtigung der natürlichen Bedingungen und wirtschaftlichen Verhältnisse.*) Munich: R. Oldenbourg, 1893. Pp. 763.

16 [22] *Anthropogeographical Contributions: on Mountain Geography with Special Reference to Upper Limits and Vertical Zoning.* (*Anthropogeographische Beiträge. Zur Gebirgskunde, vorzüglich Beobachtungen über Höhengrenzen und Höhengürtel.*) Published in the records of the Leipzig Geographical Society and of the Carl Ritter Foundation at Leipzig. Leipzig: Duncker and Humbolt, 1895. Pp. 362.

17 [23] *The History of Mankind* (Vols. I–III.) Translated from

the second German edition of *Ethnology* (1894-95) by A. J. Butler, and with an introduction by E. B. Tylor. London: Macmillan and Co., 1896-1898. Pp. 486, 562, 600.

18 [24] *Political Geography*. (*Politische Geographie*.) Munich and Leipzig: R. Oldenbourg, 1897. Pp. 715.

19 [25] *Germany: an Introduction to the Geography of the Home Country*. (*Deutschland. Einführung in die Heimatkunde*.) Leipzig: F. W. Grunow, 1898. Pp. 332.

20 [26] *Anthropogeography. Part I*. Second edition of 9. (*Anthropogeographie. 1. Teil. Grundzüge der Anwendung der Erdkunde auf die Geschichte*.) Stuttgart: J. Engelhorn, 1899. Pp. 604.

21 [27] *Contributions to the Geography of Central Germany*. (*Beiträge zur Geographie des mittleren Deutschlands*.) Published in the records of the Carl Ritter Foundation at Leipzig. Leipzig: Duncker and Humbolt, 1899. Pp. 382.

22 [28] *The Sea as a Source of National Greatness. A Study in Political Geography*. (*Das Meer als Quelle der Völkergröße. Eine politisch-geographische Studie*.) Munich: R. Oldenbourg, 1900. Pp. 86.

23 [29] *The Earth and Life. A Comparative Geography*. (Vols. I, II.) (*Die Erde und das Leben. Eine vergleichende Erdkunde*.) Leipzig: Bibliographisches Institut, 1901-1902. Pp. 706, 702.

24 [30] *Political Geography, or the Geography of States, of Trade and of War*. (*Politische Geographie oder die Geographie der Staaten, des Verkehrs und des Krieges*.) Munich and Berlin: R. Oldenbourg, 1903. (A second edition of 18 with the additional chapters on trade and war, nos. 16 and 17.) Pp. 838.

25 [31] *On Depicting Nature*. (*Über Naturschilderung*.) Munich and Berlin: R. Oldenbourg, 1904. Pp. 394.

26 [32] *Islands of Bliss and Dreams*. (*Glücksinseln und Träume*.) Collected essays from the *Grenzboten*. Leipzig: F. W. Grunow, 1905. Pp. 515.

27 [33] *Friedrich Ratzel's shorter writings*. (Vols. I, II.) (*Kleine*

Schriften von Friedrich Ratzel.) Edited by Hans Helmolt and with a bibliography by Viktor Hantzsch. Munich and Berlin: R. Oldenbourg, 1906. Pp. 531, 544.

II. PAPERS AND SHORTER CONTRIBUTIONS TO JOURNALS AND COLLECTIVE WORKS

1867

1 'Contributions to the Study of the Anatomy of Enchytraeus Vermicularis "Henle"' ('Beiträge zur Anatomie von Enchytraeus vermicularis "Henle"'). Leipzig: *Zeitschrift für wissenschaftliche Zoologie*, vol. XVIII, pp. 99–108.

1868

2 'On the History of the Development of Earthworms (*Lumbricus agricola* "*Hoffm.*")' ('Zur Entwicklungsgeschichte des Regenwurms'). *Ibid.* vol. XVIII, pp. 547–62.
3* 'Contributions to the Anatomical and General Study of "Oligochaetes"' ('Beiträge zur anatomischen und systematischen Kenntnis der Oligochaeten'). *Ibid.* vol. XVIII, pp. 563–91.
4 [6] *Zoological Letters from the Mediterranean.* (*Zoologische Briefe vom Mittelmeer*). Essays in the *Kölnische Zeitung*.

1870

5 [10] 'Protoplasm' ('Das Protoplasma'). Meyer's *Ergänzungsblätter zur Kenntnis der Gegenwart*, vol. I, pp. 679–701.
6 [11] 'Fresh progress in Zoology' ('Neuere Fortschritte der Zoologie'). *Ibid.* vol. I, pp. 762–9.

7 [12] 'Tertiary Man' ('Der tertiäre Mensch'). *Ibid.* vol. I, pp. 775–7.
8 [13] 'Recent Researches on Blood Corpuscles' ('Neuere Untersuchungen über die Blutkörperchen'). *Ibid.* vol. II, pp. 40–4.
9 [14] 'The Investigations about Animal Life in the Deep Seas' ('Die Untersuchungen über das Tierleben in der Meerestiefe'). *Ibid.* vol. II, pp. 98–104.
10 [15] 'The Oldest Remains of Organic Life (Eozoic)' ('Die ältesten Reste organischen Lebens (Eozoon)'). *Ibid.* vol. II, pp. 107–12.
11 [16] 'A. Wallace's Contributions to the Theory of Natural Selection' ('A. Wallaces Beiträge zur Theorie der natürlichen Zuchtwahl'). *Ibid.* vol. II, pp. 160–6.
12 [17] 'The Sensory Organs of Men and Animals' ('Die Sinnesorgane der Menschen und der Tiere'). *Ibid.* vol. II, pp. 228–32, 287–90.
13 [18] 'New Investigations about Birds' Nests' ('Neue Untersuchungen über die Vogelnester'). *Ibid.* vol. II, pp. 496–500.
14 [19] 'Anthropological Societies' ('Die anthropologischen Gesellschaften'). *Globus*, vol. XVII, p. 204.

1871

15 [21] 'From Transylvania' ('Aus Siebenbürgen'). Travel articles in the *Kölnische Zeitung*.

1872

*16 [24] 'Ernst Häckel.' Meyer's *Deutsches Jahrbuch*, 1. Jahrgang, pp. 555–8.
17 [26] 'Letters from Southern Italy' ('Briefe aus Süditalien'). Essays in the *Kölnische Zeitung*.
18 [27] 'From the Alps' ('Aus den Alpen'). Travel articles in the *Kölnische Zeitung*.

1873

19 [28] 'Zoology' ('Zoologie'). Meyer's *Deutsches Jahrbuch*, 2. Jahrgang, pp. 649–64.

20 [29] 'Palaeontology' ('Paläontologie'). *Ibid.* 2. Jahrgang, pp. 664–7.

*21 [31] 'A St Gotthard Journey in Winter' ('Gotthardreise im Winter'). Travel article in the *Kölnische Zeitung*.

1876

22 [34] 'Arakan under British Government' ('Arakan unter britischer Regierung'). *Globus*, vol. XXX, pp. 284–5.

23 [35] 'On the Statistics of British Burma' ('Zur Statistik von Britisch-Birma'). *Ibid.* vol. XXX, pp. 296–7.

24 [36] 'From the Cotton States' ('Aus den Baumwollenstaaten'). *Ibid.* vol. XXX, pp. 314–18, 344–6.

25 [37] 'The Report on the Material and Moral Progress of India in 1874–1875' ('Der Bericht über den materiellen und moralischen Fortschritt Indiens in 1874/1875'). *Ibid.* vol. XXX, p. 384.

26 [38] 'Appraisal of the Chinese' ('Die Beurteilung der Chinesen'). *Österreichische Monatsschrift für den Orient*, vol. II, pp. 177–82.

1877

27 [40] 'Fresh work on the Fauna of America' ('Neuere Arbeiten über die Tierwelt Amerikas'). *Globus*, vol. XXXII, pp. 202–4.

*28 [41] 'On California' ('Über Kalifornien'). 6. und 7. *Jahresbericht der Geographischen Gesellschaft in München*, pp. 124–48.

1878

29 [44] 'A Bibliography of Anthropological Literature. Ethnology and Journeys' ('Verzeichnis der anthropologischen

Literatur: Ethnologie und Reisen'). *Archiv für Anthropologie*, vol. X, pp. 51–93.

30 [47] 'On Mountaineering' ('Zur Bergsteigerei'). *Die Gegenwart*, vol. XIV, pp. 151–4, 228–30.

31 [48] 'Fresh Work on the Fauna of America.' (See item 27 [40]). *Globus*, vol. XXXIII, pp. 7–10, 77–9.

32 [49] 'Tea cultivation in India' ('Der Teebau in Indien'). *Ibid.* vol. XXXIII, pp. 247–8.

33 [50] 'Public Education in British Burma and Assam' (Der öffentliche Unterricht in Britisch-Birma und Assam'). *Ibid.* vol. XXXIII, pp. 250–1.

*34 [51] 'Some Observations on Characteristic Features of the Tropics' ('Einige Bemerkungen über tropischen Naturcharakter'). *Ibid.* vol. XXXIII, pp. 330–4, 346–7, 360–1.

35 [52] 'Recent Investigations in Lower Colorado' ('Neuere Forschungen am untern Colorado'). *Ibid.* vol. XXXIV, pp. 118–22.

36 [53] 'Geography and Ethnology at the British Association' ('Geographisches und Ethnographisches von der British Association'). *Ibid.* vol. XXXIV, pp. 202–3.

37 [54] 'Notes on Commercial and Transport Geography' ('Notizen zur Handels- und Verkehrs-Geographie'). *Ibid.* vol. XXXIV, pp. 252–5, 267–8, 381–4.

38 [55] 'The New Commercial Centres and Trade Routes in South-East Asia' ('Die neuen Handelsplätze und Handelswege in Hinterindien'). *Österreichische Monatsschrift für den Orient*, vol. IV, pp. 81–5, 97–104, 119–25.

39 [56] 'An Appraisal of the Japanese' ('Zur Beurteilung der Japaner'). *Ibid.* vol. IV, pp. 161–5.

40 [57] 'The Assessment of Peoples' ('Die Beurteilung der Völker'). *Nord und Süd*, vol. VI, pp. 177–200.

1879

*41 [59] 'On the Occasion of the Centenary of Karl Ritter's Birth' ('Zu Karl Ritters hundertjährigem Geburtstage').

Beilage zur Allgemeinen Zeitung. (A series of five short articles.)

42 [74] 'Concerning South-East Asia' ('Hinterindisches'). *Die Gegenwart*, vol. XVI, pp. 40–4, 59–60 (signed Franz Einsiedel).

*43 [75] 'The Physiognomy of the Moon' ('Die Physiognomie des Mondes'). *Ibid.* vol. XVI, pp. 124–6 (signed Franz Einsiedel).

44 [76] 'Notes on Commercial and Transport Geography' ('Notizen zur Handels- und Verkehrs-Geographie'). *Globus*, vol. XXXV, pp. 124–7, 223–4, vol. XXXVI, pp. 206–8.

45 [77] 'Forest Statistics and Forest Protection in the United States' ('Waldstatistik und Waldschutz in den Vereinigten Staaten'). *Ibid.* vol. XXXV, pp. 360–4.

46 [78] 'The Steppe at Lake Mono' ('Die Steppe am Mono-See'). *Ibid.* vol. XXXV, pp. 378–9.

47 [79] 'Reports on Quelpart Island (Korean coast)' ('Nachrichten über die Insel Quelpart'). *Ibid.* vol. XXXV, pp. 382–3.

48 [80] 'The Development of the West of the United States' ('Die Entwicklung des Westens der Vereinigten Staaten') (unsigned). *Ibid.* vol. XXXVI, pp. 237–8.

49 [81] 'A Summary of the Most Important Events of 1878 in East and South Asia, Africa and Australia' ('Chronik der bemerkenswertesten Ereignisse des Jahres 1878 in Ost- und Südasien, Afrika und Australien') (unsigned). *Österreichische Monatsschrift für den Orient*, vol. V, pp. 11–14, 35–9.

50 [82] 'Korea, the Riu-Kiu Islands and the Two Great East Asiatic Powers' ('Korea, die (R)Liukiu-Inseln und die zwei ostasiatischen Großmächte'). *Ibid.* vol. V, pp. 189–96.

51 [83] 'Geographical Studies of Baden' ('Geographische Studien über Baden'). *Karlsruher Zeitung*, 6 July.

52 [84] 'The Central Union for Commercial Geography and for the furthering of German interests overseas' ('Der Zentralverein für Handelsgeographie und Förderung deutscher Interessen im Auslande'). *Kölnische Zeitung*, no. 114 (signed F.R.).

53 [85] 'Extracts from the German Consuls' Reports for 1877 and 1878' ('Aus den Berichten der deutschen Konsuln für 1877 und 1878'). Published in the *Kölnische Zeitung* (nine articles).

1880

54 [86] 'The Inter-Ocean Canal through Central America' ('Der interozeanische Kanal durch Mittelamerika'). *Beilage zur Allgemeinen Zeitung* (five supplementary articles).

55 [87] 'An Augsburg polar explorer' (Johann Georg Karl (or Karl Ludwig) Metzler-Giesecke) ('Ein Augsburger Polarforscher'). *Beilage zur Allgemeinen Zeitung*, no. 335, pp. 4921–2) (signed R).

56 [99] 'The Kurile Islands' ('Die Kurilen'). *Globus*, vol. XXXVII, pp. 142–4 (unsigned).

57 [100] 'North America's useful plants and animals' ('Nordamerikas nutzbare Pflanzen und Tiere'). *Ibid*. vol. XXXVII, pp. 153–5, 170–4.

58 [101] 'Exploration of America since 1870' ('Die Erforschung Amerikas seit 1870'). Meyer's *Deutsches Jahrbuch*, vol. 1879–80, pp. 278–95.

*59 [103] 'Concerning the Formation of Earth-Pillars' ('Über die Entstehung der Erdpyramiden'). *Jahresbericht der Geographischen Gesellschaft in München*, 1877–79, pp. 77–88.

60 [106] 'America. Geographical and Ethnographical Investigations since 1870' ('Amerika. Geographische und ethnographische Forschungen seit 1870'). Meyer's *Konversationslexikon*, vol. XVII, pp. 28–35 (unsigned).

61 [107] 'South-East Asia. New Trade Routes and Commercial Centres' ('Hinterindien, Neue Handelswege und Handelsplätze'). *Ibid*. vol. XVII, pp. 450–1 (unsigned).

62 [108] 'Mexico: recent history' ('Mejiko, neueste Geschichte'). *Ibid*. vol. XVII, pp. 581–2 (unsigned).

63 [109] 'The Freeing of the American Slaves' ('Sklavenbefreiung in Amerika'). *Ibid*. vol. XVII, pp. 812–16 (unsigned).

*64 [110] 'A Historical comment on the concept "Mediterranean"' ('Historische Notiz zu dem Begriff "Mittelmeer"'). Petermann's *Mitteilungen*, vol. XXVI, pp. 338–40.

*65 [111] 'Concerning the Formation of Fjords on Inland Seas' ('Über Fjordbildungen an Binnenseen'). *Ibid.* vol. XXVI, pp. 387–96.

66 [112] 'Alpine Studies' ('Hochgebirgsstudien'). Westermann's *Illustrierte deutsche Monatshefte*, vol. XLVIII, pp. 374–83, 499–517, 739–53.

67 [113] 'A Summary of the Most Important Events of 1879 in East and South Asia, Africa and Australia.' (See item 49[81].) *Österreichische Monatsschrift für den Orient*, pp. 34–8, 53–6, 69–72, 90–2 (unsigned).

68 [114] 'The Chinese in North America since 1875' ('Die Chinesen in Nordamerika seit 1875'). *Ibid.* vol. VI, pp. 189–94.

*69 [115] 'Waterfalls' ('Die Wasserfälle'). *Nord und Süd*, vol. XIV, pp. 218–43.

70 [116] 'The Future and Evaluation of the Negroes' ('Zukunft und Beurteilung der Neger'). *Deutsche Revue*, vol. II, pp. 97–111.

*71 [117] 'Concerning the Geographical Conditions and Ethnographical Consequences of Human Migrations' ('Über geographische Bedingungen und ethnographische Folgen der Völkerwanderungen'). *Verhandlungen der Gesellschaft für Erdkunde zu Berlin*, vol. VII, pp. 295–324.

72 [118] 'The advance of the United States in the South American Commercial Sphere' ('Das Vordringen der Vereinigten Staaten in das südamerikanische Handelsgebiet'). *Weserzeitung*, January.

73 [119] 'A Good Goal for German Emigration' ('Ein gutes Ziel für deutsche Auswanderung'). *Allgemeine Zeitung* (signed F.R.).

1881

74 [120] 'The Opening up of Korea' ('Koreas Erschließung').
Beilage zur Allgemeinen Zeitung.

75 [121] 'The German Universities in the United States' ('Die deutsche Hochschule in den Vereinigten Staaten'). *Ibid.*

76 [127] 'Chinese Emigration since 1875' ('Die chinesische Auswanderung seit 1875'). *Globus*, vols. XXXIX and XL (fourteen contributions).

77 [128] 'The Geographical Exploration of the Americas' ('Geographische Erforschung Amerikas'). Meyer's *Konversationslexikon*, vol. II, supplementary volume, pp. 26–30 (unsigned).

78 [129] 'The Neutral Territory between China and Korea' ('Das neutrale Gebiet zwischen China und Korea'). Petermann's *Mitteilungen*, vol. XXVII, pp. 71–2.

79 [130] 'A Summary of the Most Important Events of 1880 in East and South Asia', etc. (See item 49[81].) *Österreichische Monatsschrift für den Orient*, pp. 31–5, 48–51, 64–8, 85–6 (unsigned).

80 [131] 'Baden People in the United States' ('Badenser in den Vereinigten Staaten'). *Karlsruher Zeitung*, 23 and 24 January.

1882

81 [134] 'The Position of Primitive Peoples in Human Society' ('Die Stellung der Naturvölker in der Menschheit'). *Das Ausland*, nos. 1, 2, 4 (three contributions) (unsigned).

82 [135] '1881 in Retrospect from the Point of View of Political and Economic Geography' ('Politisch- und wirtschaftsgeographische Rückblicke auf das Jahr 1881'). *Ibid.* nos. 1, 5, 6 (three contributions) (unsigned).

83 [136] 'German Participation in International Polar Exploration' ('Beteiligung des Deutschen Reiches an der internationalen Polarforschung'). *Ibid.* no. 3, pp. 41–5 (unsigned).

84 [137] 'The First Geographical Congress in Berlin' ('Der 1. deutsche Geographentag zu Berlin'). *Ibid.* no. 15, pp. 281–5 (unsigned).

85 [138] 'A Complete Summary of the Reports on the Fate of the "Jeannette" and her Crew' ('Vollständige Zusammenstellung der Nachrichten über die Schicksale der "Jeannette" und ihrer Mannschaft'). *Ibid.* nos. 17, 18, 19, 20, 21, 22, 24, 25, 27, 36 (ten contributions) (unsigned).

86 [140] 'The Second German Geographical Congress in Halle' ('Der 2. deutsche Geographentag zu Halle'). *Ibid.* no. 20, pp. 381–6 (unsigned).

87 [141] 'The Present State of German Exploration in Africa' ('Der gegenwärtige Stand der deutschen Afrikaforschung'). *Ibid.* no. 32, pp. 621–6 (unsigned).

88 [142] 'Matteucci's and Massari's Journey across Africa' ('Matteuccis und Massaris Reise quer durch Afrika'). *Ibid.* nos. 34, 38, 40 (unsigned).

89 [143] 'On the Theory of "Spheres of Notion"' ('Zur Lehre von den Ideenkreisen'). *Ibid.* no. 39, pp. 778–9 (unsigned).

90 [144] 'A Summons to Co-operation in Producing a General Geography of Germany' ('Aufruf zur Mitarbeit an einer allgemeinen deutschen Landeskunde'). *Ibid.* no. 40, pp. 781–2.

91 [145] 'Slavery and Emancipation in Cuba' ('Sklaverei und Emanzipation auf Cuba'). *Ibid.* nos. 51, 52 (short contributions) (unsigned).

1883

92 [154] 'Political and Economic Geography in retrospect' ('Politisch- und wirtschaftsgeographische Rückblicke'). *Ibid.* nos. 1, 2, 5, 13, 14, 15, 18 (seven short contributions) (unsigned). (See also item 82[135].)

93 [155] 'The First Report of the Central Committee for the Geography of Germany, with a Supplement' ('Erster Bericht des Zentralausschusses für deutsche Landeskunde nebst Beilage'). *Ibid.* no. 2, pp. 21–3.

94 [156] 'Considerations of the Nature and Exploration of the Polar regions' ('Betrachtungen über Natur und Erforschung der Polarregionen'). *Ibid.* nos. 11, 12, 13, 18, 19 (six short contributions) (unsigned).

95 [157] 'The Second Report of the Central Committee', etc. (See item 93[155].) *Ibid.* no. 13, pp. 241–2.

96 [158] 'The Third German Geographical Congress in Frankfurt a. M.' ('Der 3. deutsche Geographentag in Frankfurt a. M.'). *Ibid.* nos. 17, 18 (unsigned).

97 [159] 'Additional Comments on and Sequel to the "Jeannette" Expedition' ('Nachträge und Nachspiel der "Jeannette" Expedition'). *Ibid.* no. 22 (unsigned).

98 [167] 'The importance of Polar Investigation for Geography' ('Die Bedeutung der Polarforschung für die Geographie'). *Verhandlungen des 3. deutschen Geographentages zu Frankfurt a. M., 1883*, Berlin, 1883, pp. 21–37.

1884

99 [169] 'Considerations of the Nature and Exploration of the Polar Regions.' (See item 94[156].) *Das Ausland*, nos. 8, 11.

100 [171] 'The Fourth German Geographical Congress in Munich' ('Der 4. deutsche Geographentag in München'). *Ibid.* nos. 17, 18, 19, 21 (unsigned).

101 [172] 'Political and Economic Geography in Retrospect.' (See items 82[135] and 92[154].) *Ibid.* nos. 20, 22, 23, 26, 30 (unsigned).

102 [173] 'A Commercial Museum for Munich' ('Ein Handelsmuseum für München'). *2. Beilage zur Allgemeinen Zeitung*, no. 337, p. 1.

103 [178] 'On the Actual Position of Polar Discovery' ('Über den gegenwärtigen Stand der Polarforschung'). *Deutsche Rundschau*, vol. XXXVIII, pp. 256–77.

104 [179] 'Transactions of the Fourth German Geographical Congress at Munich in 1884, edited by the commission of the Central Committee for German Geographical Congresses by

F. Ratzel' ('Verhandlungen des 4. deutschen Geographentages zu München 1884, im Auftrage des Zentralausschusses des deutschen Geographentages herausgegeben von F. Ratzel'). Vol. IV. Berlin: D. Reimer, 1884. Pp. 191.

105 [181] 'Report of the Central Commission for the Scientific Geography of Germany' ('Bericht der Zentralkommission für wissenschaftliche Landeskunde von Deutschland'). *Verhandlungen des 4. deutschen Geographentages zu München 1884*, Berlin, 1884, pp. 141–8.

1885

106 [189] 'On the Achievements and Aims of Polar Exploration' ('Über Ergebnisse und Ziele der Polarforschung'). *Jahresbericht der Geograph. Gesellschaft in München für 1884*, pp. 21–2.

107 [190] 'The Effect of Developments in Africa on the Trend of Activity of Colonial Society' ('In welcher Richtung beeinflussen die afrikanischen Ereignisse die Tätigkeit des Kolonialvereins?'). *Deutsche Kolonialzeitung*, vol. II, pp. 38–44.

108 [191] 'The Project of a New Political Map of Africa. To Include some General Observations on the Principles of Political Geography' ('Entwurf einer neuen politischen Karte von Afrika. Nebst einigen allgemeinen Bemerkungen über die Grundsätze der politischen Geographie'). *Petermann's Mitteilungen*, vol. XXXI, pp. 245–50.

109 [192] 'Tasks of Geographical Exploration in the Antarctic' ('Aufgaben geographischer Forschung in der Antarktis'). *Verhandlungen des 5. deutschen Geographentages zu Hamburg 1885*, Berlin, 1885, pp. 8–24.

1886

110 [193] 'Through War to Peace. Sketches and Impressions of the years 1870–1871 by Karl Stieler, with an introduction by Friedrich Ratzel' ('Durch Krieg zum Frieden. Stimmungs-

bilder aus den Jahren 1870–1871 von Karl Steiler. Mit einem Vorwort von Friedrich Ratzel'). Stuttgart, Bonz und Comp., pp. 270.

111 [194] 'Emin Bey.' *Beilage zur Allgemeinen Zeitung*, no. 361.

112 [204] 'Concerning the Snow Conditions in the Bavarian Limestone Alps' ('Über die Schneeverhältnisse in den bayerischen Kalkalpen'). *Jahresbericht der Geographischen Gesellschaft in München für 1885*, pp. 24–34.

*113 [205] 'A Criticism of the So-Called "Snow-Line"' ('Zur Kritik der sogenannten "Schneegrenze"'). *Leopoldina*, vol. XXII, pp. 186–8, 201–4, 210–12.

*114 [206] 'On Photographs of Alpine Landscapes' ('Über Photographien alpiner Landschaften'). *Mitteilungen des Deutschen und Österreichischen Alpenvereins*, new series, vol. II, p. 43.

115 [207] 'A List of Queries on Snow Conditions in Mountain Regions' ('Fragebogen über die Schneeverhältnisse in Gebirgen'). *Ibid.* new series, vol. II, pp. 137–8.

116 [208] 'A New Special Map of Africa' ('Eine neue Spezialkarte von Afrika'). Petermann's *Mitteilungen*, vol. XXXII, pp. 161–2.

117 [209] 'A List of Queries on Snow Conditions' ('Fragebogen über Schneeverhältnisse'). *Ibid.* vol. XXXII, pp. 182–3.

*118 [210] 'The Determination of the Snow Line' ('Die Bestimmung der Schneegrenze'). *Der Naturforscher*, 19th annual issue, pp. 245–8.

*119 [211] 'The Geographical Picture of Mankind. A Centenary account' ('Das geographische Bild der Menschheit. Eine Centennialbetrachtung'). *Deutsche Rundschau*, vol. XLVIII, pp. 40–63.

120 [212] 'Wendelstein' (Bavarian Meteorological Station) ('Der Wendelstein'). *Zeitschrift des Deutschen und Österreichischen Alpenvereins*, vol. XVII, pp. 361–439.

BIBLIOGRAPHY

1887

121 [213] 'Report on Hans Meyer's Ascent of Kilimanjaro' ('Bericht über Hans Meyers Kilimandscharo-Besteigung'). *Beilage zur Allgemeinen Zeitung*, no. 303 (signed R.).

122 [214] 'The Geographical Distribution of Bows and Arrows in Africa' ('Die geographische Verbreitung des Bogens und der Pfeile in Afrika'). *Berichte über die Verhandlungen der Königlich Sächsischen Gesellschaft der Wissenschaften zu Leipzig, philologisch-historische Klasse*, vol. XXXIX, pp. 233–52.

*123 [222] 'Oskar Peschel.' *Allgemeine Deutsche Biographie*, vol. XXV, pp. 416–30.

*124 [225] 'An Assessment of Anthropophagy' ('Zur Beurteilung der Anthropophagie'). *Mitteilungen der Anthropologischen Gesellschaft in Wien*, vol. XVII, pp. 81–5.

*125 [226] 'The Influence of Firn on the Deposition of Debris and the Formation of Humus' ('Der Einfluß des Firnes auf Schuttlagerung und Humusbildung'). *Mitteilungen des Deutschen und Österreichischen Alpenvereins*, new series, vol. III, pp. 97–100.

126 [227] 'On Slatted Armour, and its Distribution in the North Pacific Region' ('Über die Stäbchenpanzer und ihre Verbreitung im nord-pazifischen Gebiet'). *Sitzungsberichte der philosophisch-philologischen und historischen Klasse der Kgl. Bayer. Akademie der Wissenschaften zu München, Jahrgang 1886*, pp. 181–216.

127 [228] 'New Letters from Emin Pasha' ('Neue Briefe von Emin Pascha'). *Allgemeine Zeitung*, no. 253 (signed F. R.).

1888

128 [229] 'The Northern Limit of the Boomerang in Australia' ('Die Nordgrenze des Bumerang in Australien'). *Internationales Archiv für Ethnographie*, vol. I, p. 27.

129 [230] 'The Letters and Reports of Emin Pasha' ('Die

Briefe und Berichte Emin Paschas'). *Beilage zur Allgemeinen Zeitung*, no. 81 (unsigned).

130 [231] 'On the Application of the Concept of the "Oecumene" to Contemporary Geographical Problems' (with a map) ('Über die Anwendung des Begriffs Oekumene auf geographische Probleme der Gegenwart'). *Berichte über die Verhandlungen der Königlich Sächsischen Gesellschaft der Wissenschaften zu Leipzig, philologisch-historische Klasse*, vol. XL, pp. 137–80.

*131 [234] 'Eduard Friedrich Pöppig.' *Allgemeine Deutsche Biographie*, vol. XXVI, pp. 421–7.

132 [242] 'A New Aspect of the World' ('Ein neues Erdbild'). *Die Grenzboten*, 47. Jahrgang, no. 12, pp. 587–91 (unsigned).

133 [243] 'The Development of a Consciousness of Nature' ('Die Entwicklung des Naturgefühls'). *Ibid*. 47. Jahrgang, no. 19, pp. 256–62 (unsigned).

134 [244] 'Space Relationships in History' ('Die Entfernungen in der Geschichte'). *Ibid*. 47. Jahrgang, no. 37, pp. 493–501 (unsigned).

135 [245] 'New Details about Snow Stratification' ('Neue Bruchstücke über Schneelagerung'). *Jahresbericht der Geographischen Gesellschaft in München für 1887*, pp. 69–79.

*136 [246] 'On the Art of Depicting Nature' ('Zur Kunst der Naturschilderung'). *Mitteilungen des Deutschen und Österreichischen Alpenvereins*, new series, vol. IV, pp. 161–5, 173–5.

*137 [248] 'Concerning Political Conditions in the African Interior' ('Über politische Verhältnisse in Innerafrika'). *Unsere Zeit*, vol. I, pp. 361–73.

138 [249] 'From Usambara' ('Aus Usambara'). *Allgemeine Zeitung*, no. 327.

1889

139 [250] 'A Note on Hans Meyer's and Purtscheller's Ascent of Kilimanjaro' ('Notiz über Hans Meyers und Purtschellers Besteigung des Kilimandscharo'). *Beilage zur Allgemeinen Zeitung*, no. 346.

*140 [251] 'On Anthropogeographical Concepts of the Vertical Perspective of History and the Vertical Perspective of Mankind' ('Über die anthropogeographischen Begriffe Geschichtliche Tiefe und Tiefe der Menschheit'). *Berichte über die Verhandlungen der Königlich Sächsischen Gesellschaft zu Leipzig*, etc. (See item 130[231].) Vol. XLI, pp. 301–24.

141 [254] 'Karl Ritter.' *Allgemeine Deutsche Biographie*, vol. XXVIII, pp. 679–97.

142 [256] 'William of Rubruck' ('Wilhelm von Rubruk'). *Ibid.* vol. XXIX, pp. 432–4.

143 [257] 'Perforated Stones in Chile' ('Durchbohrte Steine in Chile'). *Globus*, vol. LVI, p. 110.

144 [258] 'On Debris Formed by Ice and Firn' ('Über Eis- und Firnschutt'). Petermann's *Mitteilungen*, vol. XXXV, pp. 174–6.

*145 [259] 'Firn Patches' ('Firnflecken'). *Münchner Neueste Nachrichten*.

*146 [260] 'On Hoarfrost on the Ground' ('Über Bodenreif'). *Das Wetter*, vol. VI, no. 9, p. 216.

*147 [261] 'On Alpine Limits and Alpine Zones' ('Höhengrenzen und Höhengürtel'). *Zeitschrift des Deutschen und Österreichischen Alpenvereins*, vol. XX, pp. 102–35.

*148 [262] 'On the Measurement of the Density of Snow' ('Über Messung der Dichtigkeit des Schnees'). *Meteorologische Zeitschrift*, vol. VI, pp. 433–5.

149 [263] 'Dr Hans Meyer's Further Surveys in the Kilimanjaro Region' ('Dr. Hans Meyers weitere Aufnahmen im Kilimandscharogebiet'). *Allgemeine Zeitung*, no. 362.

1890

*150 [269] 'Avalanches in the Giant Mountains' (Bohemian Massif) ('Lawinen im Riesengebirge'). Petermann's *Mitteilungen*, vol. XXXVI, pp. 199–200.

151 [270] 'An Attempt to Sum up the Scientific Results of Stanley's [trans-African] Journey' ('Versuch einer Zusam-

menfassung der wissenschaftlichen Ergebnisse der Stanleyschen Durchquerung'). *Ibid.* vol. XXXVI, pp. 257–62, 281–96.

152 [272] 'Concerning the "Karrenfelder" (grike areas) in the Alps' ('Über Karrenfelder in den Alpen'). *Veröffentlichungen der Sektion Leipzig des Deutschen und Österreichischen Alpenvereins*, vol. V.

1891

153 [273] 'African Bows: their Distribution and Relationships. With a Supplement: Concerning the Bows of New Guinea, of the Veddah and of the Negritos. An Anthropogeographical Study' ('Die afrikanischen Bögen, ihre Verbreitung und Verwandtschaften. Nebst einem Anhange: Über die Bögen Neu-Guineas, der Veddah und der Negritos. Eine anthropogeographische Studie'). Leipzig: S. Hirzel, 1891. *Abhandlungen der philologisch-historischen Klasse der Kgl. Sächsischen Gesellschaft der Wissenschaften zu Leipzig*, vol. XIII, pp. 291–346.

154 [274] 'Sir Thomas Elder's Expedition to Central Australia' ('Die Expedition Sir Thomas Elders nach Zentralaustralien'). *Beilage zur Allgemeinen Zeitung*, nos. 127, 287 (signed R.).

155 [275] 'The Three Scientific Expeditions to Victoria Nyanza' ('Die drei wissenschaftlichen Expeditionen nach dem Viktoria-Nyanza'). *Ibid.* no. 239, p. 7 (signed R.).

156 [276] 'Exploration of Victoria Nyanza' ('Die Erforschung des Viktoria-Nyanza'). *Ibid.* no. 246, pp. 1–3.

157 [289] 'Casati and Emin Pasha' ('Casati und Emin Pascha'). *Die Grenzboten*, 50. Jahrgang, no. 10, pp. 433–43.

*158 [290] 'Concerning some Obscure Points in Glaciology' ('Über einige dunkle Punkte der Gletscherkunde'). *Veröffentlichungen der Sektion Leipzig des Deutschen und Österreichischen Alpenvereins*, vol. VI.

159 [291] 'A Reply to Hermann Wagner's Review of Anthropogeography II' ('Erwiderung auf Hermann Wagners Besprechung der Anthropogeographie II'). *Zeitschrift der Gesellschaft für Erdkunde zu Berlin*, vol. XXVI, pp. 508–12.

BIBLIOGRAPHY

160 [292] 'The German Share in the Exploration of Africa' ('Deutschlands Anteil an der Erforschung Afrikas'). *Zeitschrift für Schulgeographie*, vol. XII, pp. 150–3.

1892

161 [293] 'On the General Characteristics of Geographical Frontiers and on Political Frontiers' ('Über allgemeine Eigenschaften der geographischen Grenzen und über die politische Grenzen'). *Berichte über die Verhandlungen der Königlich Sächsischen Gesellschaft der Wissenschaften zu Leipzig, philologisch-historische Klasse*, vol. XLIV, pp. 53–104.

*162 [299] 'Concerning the Karrenfelder in the Jura and Kindred Regions' ('Über Karrenfelder im Jura und Verwandtes'). *Leipziger Dekanatsprogramm*. Ex ordinis philosophorum mandato renuntiantur philosophiae doctores...decano Friderico Ratzel..., pp. 3–26.

163 [300] 'An Appraisal of the Negroes' ('Zur Beurteilung der Neger'). *Die Grenzboten*, 51. Jahrgang, no. 1, pp. 20–4 (unsigned).

164 [301] 'The Prospects for our South-West African Protectorate' ('Die Aussichten unseres südwestafrikanischen Schutzgebietes'). *Ibid.* 51. Jahrgang, no. 4, pp. 171–5 (unsigned).

165 [302] 'Don't squint!' ('Nicht schielen!'). *Ibid.* 51. Jahrgang, no. 8, pp. 411–12 (unsigned).

166 [304] 'Now, a Word of Thanks!' ('Doch einmal Dank!'). *Ibid.* 51. Jahrgang, no. 21, pp. 378–9 (unsigned).

167 [305] 'About our Good Friends, the Swiss' ('Von unsern guten Freunden, den Schweizern'). *Ibid.* 51. Jahrgang, no. 24, pp. 481–7; no. 28, p. 88 (unsigned).

168 [306] 'African Skirmishing' ('Afrikanische Gefechte'). *Ibid.* 51. Jahrgang, no. 34, p. 373 (unsigned).

169 [307] 'Unter den Linden.' *Ibid.* 51. Jahrgang, no. 34, p. 382 (unsigned).

170 [308] 'Representation in the Community of the Peoples'

('Die Repräsentation in der Gesellschaft der Völker'). *Ibid.* 51. Jahrgang, no. 36, pp. 433–43 (unsigned).

171 [309] 'For the Protection of the German Landscape' ('Zum Schutze der deutschen Landschaft'). *Ibid.* 51. Jahrgang, no. 40, pp. 31–3 (unsigned).

172 [310] 'Francophile Swiss Manners' ('Schweizerische Franzoseleien'). *Ibid.* 51. Jahrgang, no. 40, pp. 35–6 (unsigned).

173 [311] 'The Present and Future of the Transylvanian Saxons' ('Gegenwart und Zukunft der Siebenbürger Sachsen'). *Ibid.* 51. Jahrgang, no. 49, pp. 449–57 (unsigned).

174 [312] 'The Bernoud Case' ('Der Fall Bernoud'). *Ibid.* 51. Jahrgang, no. 49, pp. 491–2 (unsigned).

175 [314] 'Bouvet Island' (South Atlantic) ('Bouvetinsel'). *Brockhaus Konversationslexikon*, 14. Auflage, vol. III, p. 385 (unsigned).

176 [315] 'The Chinese Question' ('Chinesenfrage'). *Ibid.* 14. Auflage, vol. IV, pp. 218–20 (unsigned).

177 [316] 'Deception Bay, Deception Island' (Labrador) ('Deceptionbai, Deceptioninsel'). *Ibid.* 14. Auflage, vol. IV, p. 852 (unsigned).

178 [317] 'Political Frontiers' ('Die politischen Grenzen'). *Mitteilungen der Geographischen Gesellschaft für Thüringen zu Jena*, vol. XI, pp. 69–73.

179 [318] 'Dr. Emin Pascha.' *Deutsche Revue*, 17. Jahrgang, vol. II, pp. 211–23.

180 [319] 'On the Cartographical Presentation of Population Density and Distribution' ('Über kartographische Darstellung der Bevölkerungsdichtigkeit und Verteilung'). *Verhandlungen des 5. internationalen Kongresses der geographischen Wissenschaften zu Bern, 1891*, pp. 541.

1893

181 [321] 'A Contribution to the Knowledge of the Distribution of the Bow and the Spear in Indo-African Societies' ('Beiträge zur Kenntnis der Verbreitung des Bogens und des

Speeres im indo-afrikanischen Völkerkreis'). *Berichte über die Verhandlungen der Königlich Sächsischen Gesellschaft der Wissenschaften zu Leipzig, philologisch-historische Klasse*, vol. XLV, pp. 147–82.

182 [333] 'The Political Situation in the Hawaiian Islands' ('Die politische Lage auf den hawaiischen Inseln'). *Die Grenzboten*, 52. Jahrgang, no. 8, pp. 353–9 (unsigned).

183 [334] 'Hawaii.' *Ibid.* 52. Jahrgang, no. 10, pp. 499–501 (unsigned).

184 [335] 'Where are the Clouds?' ('Wo steh'n die Wolken?'). *Ibid.* 52. Jahrgang, no. 25, pp. 565–9 (unsigned).

185 [336] 'A Perverted Cult of Bismarck' ('Verkehrter Bismarckkultus'). *Ibid.* 52. Jahrgang, no. 29, pp. 141–2 (unsigned).

186 [340] 'Germany and the Mediterranean' ('Deutschland und das Mittelmeer'). *Ibid.* 52. Jahrgang, no. 44, pp. 196–206 (unsigned).

187 [341] 'Germany and France' ('Deutschland und Frankreich'). *Ibid.* 52. Jahrgang, no. 46, pp. 289–94 (unsigned).

188 [342] 'Ethnography and Ethnology' ('Ethnographie und Ethnologie'). *Brockhaus' Konversationslexikon*, 14. Auflage, vol. VI, pp. 389–90 (unsigned).

189 [343] 'Europa'. *Ibid.* 14. Auflage, vol. VI, pp. 422–31 (unsigned).

190 [344] 'Firn.' *Ibid.* 14. Auflage, vol. VI, pp. 823 (unsigned).

191 [345] 'Glacier' ('Gletscher'). *Ibid.* 14. Auflage, vol. VIII, pp. 71–2 (unsigned).

192 [347] 'Snow, Firn and Irrigation in the North American West' ('Schnee, Firn und Bewässerung im nordamerikanischen Westen'). Petermann's *Mitteilungen*, vol. XXXIX, pp. 20–2.

193 [348] 'Political Frontiers' ('Die politische Grenze'). *Zeitschrift für Schulgeographie*, vol. XIV, pp. 135–9.

1894

*194 [349] 'Lewis Morgan's Investigations into the Development of the State' ('Lewis Morgans Forschungen über die Entwicklung des Staats'). *Beilage zur Allgemeinen Zeitung*, no. 208, pp. 1–3; no. 209, pp. 1–4.

*195 [354] 'Concerning Geographical Position. A Political-Geographical Consideration' ('Über die geographische Lage. Eine politisch-geographische Betrachtung'). *Feestbundel aan Dr. P. J. Veth aangeboden*, Leiden, pp. 257–61.

196 [355] 'On the Development of Coasts. Anthropogeographical fragments' ('Zur Küstenentwicklung. Anthropogeographische Fragmente'). *Festschrift der Geographischen Gesellschaft in München zur Feier ihres 25-jährigen Bestehens*, pp. 67–90.

197 [357] 'Schäffle on the Eastern Question' ('Schäffle über die orientalische Frage'). *Die Grenzboten*, 53. Jahrgang, no. 2, pp. 57–61 (unsigned).

198 [359] 'People and Spaces' ('Völker und Räume'). *Ibid.* 53. Jahrgang, no. 14, pp. 1–10 (unsigned).

199 [362] 'The Deterioration of Necrology' ('Der Verfall der Nekrologie'). *Ibid.* 53. Jahrgang, no. 18, pp. 238–9 (unsigned).

200 [363] 'Off with the Mask!' ('Die Maske ab!'). *Ibid.* 53. Jahrgang, no. 21, pp. 337–40 (unsigned).

201 [364] 'English Hypocrisy again' ('Nochmals die englische Heuchelei'). *Ibid.* 53. Jahrgang, no. 23, pp. 472–3 (unsigned).

202 [367] 'German East Africa in a Clearer Light' ('Deutschostafrika in hellerm Lichte'). *Ibid.* 53. Jahrgang, no. 43, pp. 167–77 (unsigned).

203 [368] 'The Anti-Semitism of the Eighteen-Sixties' ('"1860-er" Antisemitismus'). *Ibid.* 53. Jahrgang, no. 45, pp. 282–3 (unsigned).

204 [369] 'Upper Limits of Vertical Zoning' ('Höhengrenzen').

Brockhaus' Konversationslexikon, 14. Auflage, vol. IX, pp. 264–5 (unsigned).

205 [372] 'Snow and Ice in South China in January 1893' ('Schnee und Eis in Südchina im Januar 1893'). Petermann's *Mitteilungen*, vol. XL, pp. 17–19.

206 [373] 'Geographical Observations on the Panama Affair' ('Geographische Bemerkungen zur Panamaangelegenheit'). *Mitteilungen des Vereins für Erdkunde zu Leipzig*, 1893, pp. xx–xxi.

1895

207 [374] 'Nature and Men on the Moluccas' ('Natur und Menschen auf den Molukken'). *Beilage zur Allgemeinen Zeitung*, no. 58, pp. 1–5.

*208 [375] 'Island People and Island States. A Study in Political Geography' ('Inselvölker und Inselstaaten. Eine politisch-geographische Studie'). *Ibid*. no. 301, pp. 1–4; no. 302, pp. 3–6.

209 [377] 'Bernhard Varenius.' *Allgemeine Deutsche Biographie*, vol. XXXIX, pp. 487–90.

210 [380] 'The Madagascans' ('Die Madagassen'). *Gaea*, vol. XXXI, pp. 26–39.

211 [381] 'An Examination of English World Policy' ('Zur Kenntnis der englischen Weltpolitik'). *Die Grenzboten* (ten contributions: unsigned).

212 [383] 'The Symptoms of the Cuban Revolt' ('Die Zeichen des cubanischen Aufstandes'). *Ibid*. 54. Jahrgang, no. 36, pp. 479–81 (unsigned).

213 [385] 'From German America' ('Aus Deutschamerika'). *Ibid*. 54. Jahrgang, no. 47, pp. 403–4 (unsigned).

214 [386] 'The Dardanelles and the Nile' ('Dardanellen und Nil'). *Ibid*. 54. Jahrgang, no. 50, pp. 513–22; no. 51, pp. 561–8 (unsigned).

215 [387] 'Studies on Political Regions' ('Studien über politische Räume'). *Geographische Zeitschrift*, vol. I, pp. 163–82, 286–302.

216 [388] '"Long" houses as Individual Dwellings on Seran' ('Langhäuser als Einzelwohnungen auf Seran'). *Ibid.* vol. I, p. 347.

*217 [389] 'East Asia and the United States' ('Ostasien und die Vereinigten Staaten'). *Leipziger Zeitung*, no. 93, p. 1394 (signed R.).

1896

218 [390] 'The State and its Land Geographically Considered' ('Der Staat und sein Boden geographisch betrachtet'). *Abhandlungen der philologischen Klasse der Königlichen Sächsischen Gesellschaft der Wissenschaften zu Leipzig*, vol. XVII, no. 4, pp. 127, Leipzig, S. Hirzel.

219 [391] 'Dr Hermann Meyer's Discovery Expedition' ('Über die Forschungsexpedition des Dr. Hermann Meyer'). *Beilage zur Allgemeinen Zeitung*, no. 131, pp. 6–7.

*220 [392] 'Science and Public Education in Germany' ('Wissenschaft und Volksbildung in Deutschland'). *Ibid.* no. 236, pp. 1–4; no. 237, pp. 2–5.

221 [393] 'Sea-Power: a Political-Geographical Study' ('Die Seemacht. Eine politisch-geographische Studie'). *Wissenschaftliche Beilage zur Leipziger Zeitung*, nos. 123–4, pp. 489–92 and 493–5 (signed R.).

*222 [397] 'Moritz Wagner.' *Allgemeine Deutsche Biographie*, vol. XL, pp. 532–43.

*223 [398] 'Concerning Edward Vogel's Death in Wadaï'(Lake Chad territory) ('Über den Tod Eduard Vogels in Wadaï'). *Biographische Blätter*, vol. II, pp. 45–9.

224 [399] 'Our Duty in the Transvaal Affair' ('Unsere Pflicht in Transvaal'). *Die Grenzboten*, 55. Jahrgang, no. 2, pp. 83–6 (unsigned).

225 [400] 'The Transvaal Affair' ('Zur Transvaalangelegenheit'). *Ibid.* 55. Jahrgang, no. 4, pp. 202–3 (unsigned).

226 [401] 'What Can Germany Gain from the Extension of University Education?' ('Was kann Deutschland aus der

Ausdehnung des Hochschulunterrichts gewinnen?'). *Ibid.* 55. Jahrgang, no. 22, pp. 409–23 (unsigned).

227 [402] 'Press Rumours and German Colonial Gossip' ('Deutscher Kolonien- und Zeitungsklatsch'). *Ibid.* 55. Jahrgang, no. 24, p. 527 (unsigned).

228 [403] 'German-Chinese' ('Deutsch-Chinesisch'). *Ibid.* 55. Jahrgang, no. 26, pp. 622–3 (unsigned).

229 [404] 'Our National Costumes' ('Unsere Volkstrachten'). *Ibid.* 55. Jahrgang, no. 34, pp. 357–64 (unsigned).

230 [405] 'Germany's Geographical Position ('Die geographische Lage Deutschlands'). *Ibid.* 55. Jahrgang, no. 35, pp. 390–7; no. 36, pp. 450–6 (unsigned).

231 [406] 'Germany's Position' ('Deutschlands Lage'). *Ibid.* 55. Jahrgang, no. 42, pp. 105–9 (unsigned).

232 [407] 'The State as an Organism' ('Der Staat als Organismus'). *Ibid.* 55. Jahrgang, no. 52, pp. 614–23 (unsigned).

*233 [408] 'The German Landscape' ('Die deutsche Landschaft'). Half-yearly number of the *Deutsche Rundschau*, Jahrgang 1895/6, vol. IV, pp. 407–28.

234 [409] 'The Territorial Growth of States.' *The Scottish Geographical Magazine*, vol. XII, pp. 351–61. (Abstract of 235 [410].

235 [410] 'The Laws of the Spatial Growth of States, a Contribution to Scientific Political Geography' ('Die Gesetze des räumlichen Wachstums der Staaten. Ein Beitrag zur wissenschaftlichen politischen Geographie'). Petermann's *Mitteilungen*, vol. XLII, pp. 97–107.

*236 [411] 'The Alps and their Part in Historical Movements' ('Die Alpen inmitten der geschichtlichen Bewegungen'). *Zeitschrift des Deutschen und Österreichischen Alpenvereins*, vol. XXVII, pp. 62–88.

1897

237 [413] 'The Nicaragua Canal and the Monroe Doctrine' ('Der Nicaraguakanal und die Monroe-Doktrin'). Supplement to the *Allgemeine Zeitung*, no. 227, pp. 1–7.

238 [415] 'The Greek Question' ('Die griechische Frage'). *Die Gegenwart*, vol. LII, no. 40, p. 212.

239 [416] 'Travel Sketches' ('Reiseschilderungen'). *Die Grenzboten*, 56. Jahrgang, no. 13, pp. 643–8 (unsigned).

240 [417] 'Culture' ('Bildung'). *Ibid.* 56. Jahrgang, no. 15, pp. 110–12 (unsigned).

241 [418] 'Dr. Karl Peters'. *Ibid.* 56. Jahrgang, no. 18, pp. 252–6 (unsigned).

242 [420] 'Germany's Position and Rights on the Niger' ('Deutschlands Stellung und Rechte am Niger'). *Ibid.* 56. Jahrgang, no. 25, pp. 545–51 (unsigned).

243 [421] 'Vaud Canton and "Reich's" territory' ('Waadtland und Reichsland'). *Ibid.* 56. Jahrgang, no. 30, p. 192 (unsigned).

244 [422] 'On the History of "Germandom" in North America' ('Zur Geschichte des Deutschtums in Nordamerika'). *Ibid.* 56. Jahrgang, no. 37, pp. 519–22 (unsigned).

245 [423] 'Wanderings in Old Bavaria' ('Altbayerische Wanderungen'). *Ibid.* 56. Jahrgang, no. 42, pp. 134–44; no. 43, pp. 180–9; no. 44, pp. 229–37 (unsigned).

*246 [425] 'Gerhard Friedrich Rohlfs.' *Biographisches Jahrbuch und Deutscher Nekrolog*, vol. I, pp. 325–32.

247 [427] 'The Chinese Question' ('Chinesenfrage'). *Brockhaus' Konversationslexikon*, 14. Auflage, vol. XVII, Supplement, pp. 261–2 (unsigned).

*248 [428] 'The Eastern Questions' ('Die orientalischen Fragen'). *Das Leben*, vol. I, pp. 230–45.

249 [429] 'Concerning the Causes of the Death of the African Traveller, Karl Vogel' ('Über die Ursachen des Todes des Afrikareisenden Karl Vogel'). *Mitteilungen des Vereins für Erdkunde zu Leipzig*, 1896, pp. xi–xii.

250 [430] 'Concerning Living Space. A biogeographical Sketch' ('Über den Lebensraum. Eine biogeographische Skizze'). *Die Umschau*, vol. I, no. 21, pp. 363–7.

BIBLIOGRAPHY

251 [431] 'Geographical Method in Ethnography' ('Die geographische Methode in der Ethnographie'). *Geographische Zeitschrift*, vol. III, pp. 268–78.

1898

*252 [432] 'In Memory of Heinrich Noé' ('Zur Erinnerung an Heinrich Noé'). *Beilage zur Allgemeinen Zeitung*, no. 148, pp. 1–4.

253 [433] 'The German Deep Sea Expedition' ('Die deutsche Tiefsee-Expedition'). *Wissenschaftliche Beilage der Leipziger Zeitung*, no. 6, pp. 21–3.

254 [434] 'The Origin and Movement of Peoples, Geographically Considered. Part I: Introduction and Method' ('Der Ursprung und das Wandern der Völker geographisch betrachtet. I. Mitteilung: Zur Einleitung und Methodisches'). *Berichte über die Verhandlungen der Königlich Sächsischen Gesellschaft der Wissenschaften zu Leipzig, philologisch-historische Klasse*, vol. L, pp. 1–75.

255 [435] 'German–English Relations' ('Die deutsch-englischen Beziehungen'). *Die Gegenwart*, vol. LIV, no. 27, pp. 1–2.

256 [436] 'The German Village Inn: a Travel Study' ('Das deutsche Dorfwirtshaus. Eine Wanderstudie'). *Die Grenzboten*, 57. Jahrgang, no. 1, pp. 28–34; no. 2, pp. 88–96; no. 3, pp. 143–54; no. 6, pp. 298–308 (unsigned).

257 [437] 'South-West German Wanderings' ('Südwestdeutsche Wanderungen'). *Ibid.* 57. Jahrgang (four articles) (unsigned).

258 [438] 'Thoughts on the Connections between German Territory and German History' ('Betrachtungen über den Zusammenhang zwischen dem deutschen Boden und der deutschen Geschichte'). *Ibid.* 57. Jahrgang, no. 39, pp. 591–600 (unsigned).

259 [439] 'The German Historical Landscape' ('Die deutsche historische Landschaft'). *Ibid.* 57. Jahrgang, no. 44, pp. 251–9.

260 [440] 'Travel Sketches' ('Reisebeschreibungen'). Fortnightly number of the *Deutsche Rundschau*, Jahrgang 1897/8, vol. III, pp. 263-91.

*261 [442] 'The Mountain. A Study in Landscape Morphology' ('Der Berg. Eine landschaftlich-morphologische Betrachtung'). *Mitteilungen des Deutschen und Österreichischen Alpenvereins*, new series, vol. XIV, pp. 147-9, 161-3. (See also 270 [451].)

*262 [443] 'The Naval Question and World Position' ('Flottenfrage und Weltlage'). *Münchener Neueste Nachrichten*, 51. Jahrgang, no. 4.

*263 [444] 'The German Deep-Sea Expedition' ('Die deutsche Tiefsee-Expedition'). *Die Natur*, vol. XLVII, no. 8, pp. 85-8.

264 [445] 'The Soil and the Population' ('Il suolo e la popolazione'). *Rivista italiana di Sociologia*, vol. II, pp. 139-51.

*265 [446] 'Lombard Landscapes' ('Lombardische Landschaften'). *Die Umschau*, vol. II, no. 28, pp. 481-4.

*266 [447] 'Ajaccio: a Page from a Corsican Diary' ('Nach Ajaccio. Korsisches Tagebuchblatt'). *Die Zeit*, vol. XVII, no. 221, pp. 198-200.

267 [448] 'Thoughts in Retrospect on Political Geography. I. General: Central Europe with France. II. The British Empire. III. The Russian Empire' ('Politisch-geographische Rückblicke. I. Allgemeines. Mitteleuropa mit Frankreich. II. Das englische Weltreich. III. Das russische Reich'). *Geographische Zeitschrift*, vol. IV, pp. 143-56, 211-24, 268-74.

268 [449] 'Memorandum on a Session of the German South Pole Commission' ('Notiz über eine Sitzung der deutschen Südpolar-Kommission'). *Ibid.* vol. IV, pp. 173-4 (unsigned).

*269 [450] 'Ethnography and Historical Science in America. With an Appendix by K. Lamprecht' ('Ethnographie und Geschichtswissenschaft in Amerika. Mit einem Zusatz von K. Lamprecht'). *Deutsche Zeitschrift für Geschichtswissenschaft*, Neue Folge, 2. Jahrgang, Monatsblätter, no. 3-4, pp. 65-72.

270 [451] 'The Mountain. A Study in Landscape Morphology' ('Der Berg. Eine landschaftlich-morphologische Betrachtung'). *Zeitschrift für Schulgeographie*, vol. XIX, pp. 341–8. (See also 261[442].)

271 [452] 'The Philosophy of History as Sociology' ('Die Philosophie der Geschichte als Soziologie'). *Zeitschrift für Sozialwissenschaft*, vol. I, pp. 19–25.

1899

272 [453] 'La Corse. Étude anthropogéographique' (translated by M. Zimmermann). *Annales de Géographie*, vol. VIII, pp. 304–29.

*273 [454] 'Aleria (Corsica). A historic landscape' ('Aleria. Historische Landschaft'). Scientific supplement to the *Leipziger Zeitung*, no. 83, pp. 353–5.

274 [455] 'Mariana. A Corsican landscape' ('Mariana. Eine korsische Landschaft'). *Die Gegenwart*, vol. LV, no. 1, pp. 9–11.

*275 [456] 'Corsican Towns' ('Korsische Städte'). *Globus*, vol. LXXVI, pp. 1–3, 27–31.

276 [457] 'The Letters of a Returned Wanderer' ('Briefe eines Zurückgekehrten'), 1–4. *Die Grenzboten*, 58. Jahrgang, no. 34, pp. 337–43; no. 37, pp. 505–16; no. 39, pp. 592–602; no. 50, pp. 582–92 (unsigned).

*277 [458] The Life of a Black Forest Pedlar' ('Das Leben eines Schwarzwälder Hausierers'). *Der Kynast*, 1. Jahrgang, pp. 273–5.

278 [459] 'An Anthropological Study of Corsica.' *The Scottish Geographical Magazine*, vol. XV, pp. 639–46.

279 [460] 'Macchia and Woodland in Corsica' ('Macchia und Wald in Korsika'). *Die Natur*, vol. XLVIII, no. 1, pp. 4–6; no. 3, pp. 29–30.

*280 [461] 'Spring in North Italy and Corsica' ('Der Frühling in Oberitalien und Korsika'). *Ibid.* vol. XLVIII, no. 20, pp. 229–31.

*281 [462] 'The Origin of the Aryans in its Geographical Context' ('Der Ursprung der Arier in geographischem Licht'). *Die Umschau*, vol. III, no. 42, pp. 825–7; no. 43, pp. 839–41.

282 [464] 'Mankind as a Phenomenon of Life on Earth' ('Die Menschheit als Lebenserscheinung der Erde'). In *Weltgeschichte*, edited by Hans F. Helmolt, vol. I, Leipzig, Bibliographisches Institut, 1899, pp. 61–104.

283 [465] 'The Projected German South Polar Expedition' ('Die geplante deutsche Südpolarexpedition'). *Kölnische Zeitung*, no. 432 (unsigned).

1900

284 [466] 'The Origin and Movement of Peoples, Geographically Considered.' (See 254[434].) Vol. II: 'A Geographical Examination of the Actual Origins of the European Peoples' ('Geographische Prüfung der Tatsachen über den Ursprung der Völker Europas'). *Berichte über die Verhandlungen der Königlich Sächsischen Gesellschaft der Wissenschaften zu Leipzig, philologisch-historische Klasse*, vol. LII, pp. 23–147.

285 [468] 'Max Graf von Zeppelin.' *Allgemeine Deutsche Biographie*, XLV, pp. 83–4.

286 [470] 'Myths and Ideas about the Origin of Peoples' ('Mythen und Einfälle über den Ursprung der Völker'). *Globus*, vol. LXXVIII, pp. 21–5, 45–8.

287 [471] 'The Queen of the Night' ('Die Königin der Nacht'). *Die Grenzboten*, 59. Jahrgang, no. 40, pp. 31–42 (unsigned).

288 [472] (The Letters of a Returned Wanderer' ('Briefe eines Zurückgekehrten'). *Ibid.* 59. Jahrgang, no. 41, pp. 77–87 (unsigned). (See also 276[457].)

289 [474] 'La Corsica. Studio antropogeografico.' Translated by Bartolomeo Gilardi. *Rivista geografica Italiana*, vol. VII, pp. 410–18. (See item 278[459].)

290 [476] 'The Great Powers of the Future' ('Die Großmächte der Zukunft'). *Die Woche*, no. 6.

*291 [477] 'Concerning a Law of Landscape Creation and Reproduction' ('Über ein Gesetz landschaftlicher Bildung und Nachbildung'). *Die Zeit*, vol. XXIV, no. 303, pp. 39–41.

292 [478] 'Position as a Focal Question of Geographical Instruction' ('Die Lage im Mittelpunkt des geographischen Unterrichts'). *Geographische Zeitschrift*, vol. VI, pp. 20–7. (See also 302[488].)

*293 [479] 'Some problems of a Political Ethnography' ('Einige Aufgaben einer politischen Ethnographie'). *Zeitschrift für Sozialwissenschaft*, vol. III, pp. 1–19.

1901

294 [480] 'From Transylvania' ('Aus Siebenbürgen'). Supplement to the *Allgemeine Zeitung*, no. 165, pp. 1–6 (signed F.R.).

295 [481] 'Living Space: a Biogeographical Study' ('Der Lebensraum. Eine biogeographische Studie'). Complimentary publication for Albert Schäffle on the 75th anniversary of his birthday on 24 February 1901, offered by K. Bücher, K. V. Fricker, F. X. Funck, G. v. Mandry, G. v. Mayr, F. Ratzel. Tübingen: H. Laupp. (Pp. 390; pp. 103–89 Ratzel's contribution.)

296 [482] 'Letters from a Returned Wanderer' (See 276[457] and 288[472]). *Die Grenzboten*, 60. Jahrgang, no. 13, pp. 599–609; no. 48, 434–41; no. 51, pp. 589–95 (unsigned). (See also items 275 [456], 287 [471].)

297 [483] 'The Present Evaluation of Gustav Theodor Fechner' ('Die Tagesansicht Gustav Theodor Fechners'). *Ibid.* 60. Jahrgang, no. 17, pp. 169–78.

298 [484] 'Baedeker.' *Ibid.* 60. Jahrgang, no. 44, pp. 235–45.

*299 [485] 'The Kant–Laplace Hypothesis and Geography' ('Die Kant–Laplacesche Hypothese und die Geographie'). *Petermann's Mitteilungen*, vol. XLVII, pp. 217–25.

*300 [486] 'The Spirit that Moves upon the Face of the Waters' ('Der Geist, der über den Wassern schwebt'). *Deutsche Monatsschrift*, vol. I, pp. 42–58.

301 [487] 'The Origin of the Aryans in its Geographical Context' ('Der Usprung der Arier in geographischem Licht'). *Transactions of the VIIth International Geographical Congress in Berlin, 1899*, vol. II, pp. 575–85.

302 [488] 'Position as a Focal Question of Geographical Instruction.' (See 292[478].) ('Die Lage im Mittelpunkt des geographischen Unterrichtes.') *Ibid.* vol. II, pp. 931–40.

*303 [489] 'From the Fichtel Mountains' ('Aus dem Fichtelgebirge'). *Kölnische Zeitung*, no. 135.

304 [490] 'The Awareness of Nature in our Own Time' ('Das Naturgefühl unserer Zeit'). *Die Zukunft*, vol. XXXV, no. 27, pp. 7–18.

1902

*305 [491] 'Water in Landscape' ('Das Wasser in der Landschaft'). *Unser Anhaltland*, nos. 19–21. (Also in *Globus*, LXXXI.) Pp. 126–30, 143–7.

306 [492] 'The Importance of the Time Factor in the Evolutionary Sciences' ('Die Zeitforderung in den Entwicklungswissenschaften'). *Annalen der Naturphilosophie*, vol. I, pp. 309–63. (See also 314[503]).

307 [493] 'Sociological Periodicals' ('Soziologische Zeitschriften'). *Beilage zur Allgemeinen Zeitung*, no. 80, pp. 54–5.

308 [497] 'Newly Discovered Megalithic Memorials in Corsica' ('Neue megalithische Denkmäler auf Korsika'). *Globus*, vol. LXXXII, p. 162.

309 [498] 'The Evolution and Creation of the World. With an Appendix on Lyell's and Darwin's Conceptions of the Deity' ('Weltentwicklung und Weltschöpfung. Mit einem Anhang über Lyells und Darwins Gottesideen'). *Die Grenzboten*, 61. Jahrgang, no. 24, pp. 569–84 (unsigned).

310 [499] 'Clouds in Landscape' ('Die Wolken in der Land-

schaft'). *Halbmonatshefte der Deutschen Rundschau*, Jahrgang 1901/2, vol. IV, pp. 89–117.

*311 [500] 'Bruno Hassenstein.' Petermann's *Mitteilungen*, vol. XLVIII, no. 12, pp. 1–5.

*312 [501] 'Land and Landscape in the National Spirit of the North Americans' ('Land und Landschaft in der nordamerikanischen Volksseele'). *Deutsche Monatsschrift*, vol. II, pp. 523–38.

313 [502] 'The Australian Federation and New Zealand' ('Der australische Bund und Neuseeland'). *Geographische Zeitschrift*, vol. VIII, pp. 425–50, 507–34.

1903

314 [503] 'The Importance of the Time Factor in the Evolutionary Sciences' ('Die Zeitforderung in den Entwicklungswissenschaften'). *Annalen der Naturphilosophie*, vol. II, pp. 40–97. (See also 306[492].)

*315 [504] 'Lenau and Nature' ('Lenau und die Natur'). *Beilage zur Allgemeinen Zeitung*, no. 218, pp. 585–7; no. 219, pp. 595–7; no. 220, pp. 603–6.

*316 [510] 'Emin Pascha (alias Eduard Schnitzer).' *Allgemeine Deutsche Biographie*, vol. XLVIII, pp. 346–53.

*317 [511] 'Heinrich Schurtz.' *Deutsche Geographische Blätter*, vol. XXVI, pp. 51–63.

318 [512] 'Oh Friends, the Sublime does not Dwell in Space!' ('Freunde, im Raum wohnt das Erhabne nicht!'). *Glauben und Wissen*, vol. I, pp. 19–22.

319 [513] 'The Enjoyment of Nature' ('Der Naturgenuß'). *Ibid.* vol. I, pp. 317–25.

320 [514] 'A Contribution to the Study of the Origins of German Colonial Policy' ('Ein Beitrag zu den Anfängen der deutschen Kolonialpolitik'). *Die Grenzboten*, 62. Jahrgang, no. 2, pp. 115–16 (unsigned).

321 [515] 'In the Military Hospital' ('Im Lazarett'), 1–4. *Ibid.* 62. Jahrgang, nos. 16, 17, 18, 19 (four short articles).

FRIEDRICH RATZEL

*322 [516] 'The Geographical Positions of Great Cities' ('Die geographische Lage der großen Städte'). Published in *The Great City* (*Die Großstadt*), pp. 33–72. Lectures and essays on urban geography by K. Bücher, F. Ratzel, G. v. Mayr, H. Waentig, G. Simmel, Th. Petermann and D. Schäfer. Dresden: v. Zahn and Jaensch, 1903. Pp. 282.

323 [517] 'The North-Atlantic Powers. A Study in Political Geography' ('Die nordatlantischen Mächte. Eine politisch-geographische Studie'). *Marine-Rundschau*, pp. 911–39, 1047–62.

*324 [518] 'The Distant View' ('Der Fernblick'). *Mitteilungen des Deutschen und Österreichischen Alpenvereins*, Neue Folge, vol. XIX, pp. 153–5, 165–8, 189–91, 201–3.

325 [520] 'On the Completion of the Friedrich Building in Heidelberg Castle' ('Zur Vollendung des Friedrichsbaues am Heidelberger Schloß'). Baden Museum. Supplement to the *Badische Landeszeitung*.

326 [521] 'Nationalities and Races' ('Nationalitäten und Rassen'). *Türmer-Jahrbuch*, 1904, pp. 43–77.

327 [522] 'Art, Science and the Depicting of Nature' ('Kunst, Wissenschaft und Naturschilderung'). *Die Umschau*, 7. Jahrgang, nos. 41, 42, pp. 801–4, 827–31.

328 [523] 'The Description of Nature in Geography' ('Die Naturschilderung in der Geographie'). *Vierteljahrshefte für den Geographischen Unterricht*, pp. 191–203.

329 [525] 'Geographical Conditions and Principles of Transport and Sea Strategy' ('Die geographischen Bedingungen und Gesetze des Verkehrs und der Seestrategik'). *Geographische Zeitschrift*, vol. IX, pp. 489–513.

1904

*330 [526] 'Geographical Method Applied to the Question of the Original Homeland of the Indo-Germanic Peoples' ('Die geographische Methode in der Frage nach der Urheimat der Indo-Germanen'). *Archiv für Rassen- und Gesellschaftsbiologie*, vol. I, no. 3, pp. 377–85.

*331 [527] 'The Question of the Indo-Germanic Homeland' ('Zur Frage der Indogermanenheimat'). *Ibid.* vol. I, no. 4, pp. 579–80.

*332 [528] 'A Geographical View of North America. "Germandom" in North America' ('Ein geographischer Blick auf Nordamerika. Das Deutschtum in Nordamerika'). Parts VII and VIII of the introduction to K. Baedeker's *North America. The United States including an excursion to Mexico. A handbook for travellers.* Leipzig: K. Baedeker (6th ed.), pp. xxxvii–xlix, li–liii.

333 [529] 'Studies of the Coastal Belt' ('Studien über den Küstensaum'). *Berichte über die Verhandlungen der Königlich Sächsischen Gesellschaft der Wissenschaften zu Leipzig, philologisch-historische Klasse*, vol. LV, pp. 199–298.

334 [532] 'The Central European Economic Union' ('Der mitteleuropäische Wirtschaftsverein'). *Die Grenzboten*, 63. Jahrgang, no. 5, pp. 253–9.

335 [533] 'Paracelsus.' *Ibid.* 63. Jahrgang, no. 20, pp. 238–40 (unsigned).

336 [534] 'Islands of Bliss and Dreams' ('Glücksinseln und Träume'). *Ibid.* 63. Jahrgang, nos. 40, 41, 42, 43, 45, 46. (See also Part I, entry 26.)

337 [535] 'The Conception of Nature and the Understanding of Nature' ('Naturauffassung und Naturverständnis'). *Deutsche Monatsschrift*, vol. VI, pp. 232–41, 347–57.

338 [536] 'The Central Position of Germany' ('Die zentrale Lage Deutschlands'). Foreword to W. Paszkowski's *Introductory reader to the knowledge of Germany and its cultural life*. Berlin, 1904.

*339 [537] 'In a Rock Crystal' ('In einem Bergkristall'). *Deutsche Rundschau*, 30. Jahrgang, no. 4, pp. 42–56.

340 [538] 'Long-Distance Influences from the East' ('Fernwirkungen aus Osten'). *Die Woche*, no. 30, pp. 1303–6.

*341 [539] 'History, Ethnology and Historical Perspective' ('Geschichte, Völkerkunde und historische Perspektive'). *Historische Zeitschrift*, vol. XCIII, new issue LVII, pp. 1–46.

1905

342 [541] 'Sketches of the Franco-Prussian War. From Posthumous Papers' ('Bilder aus dem Deutsch-Französischen Kriege. Aus einem Nachlaß'). *Die Grenzboten*, 64. Jahrgang, nos. 1, 2, 3, 4, 5, 6 (short articles).

INDEX

Achelis, T., geographer, 4
Africa as colonial territory, 25, 26
Africans in U.S.A., 13, 14
Agassiz, J. L. R., zoologist and glaciologist, 13
Anthropogeography, 1, 23, 33, 35, 58
Asiatics in U.S.A., 13, 14

Bastian, A., anthropologist, 24
Boer War, 38
Bowman, I., geographer, 41
Brockhaus, publishers, 34
Broek, J. O. M., geographer, 23
Brunhes, J., geographer, 18, 43
Buckle, H. T., historian, 36
Buschan, G., anthropologist, 25

Cartography, 29
Chamberlain, H. S., political philosopher, 36
Chinesische Auswanderung, Die, 13, 15, 57
colonial methods, British, 13–14
colonies, German, 22–3
Comte, A., philosopher, 36, 40
Cord-Meyer, H., historian, 39

Darwin, C., 7, 8, 11, 17, 19, 20, 39
Davis, W. M., geologist, 31

Eckert, M., cartographer, 21, 27, 28, 29, 30
Emin Pasha, 37
Erde und das Leben, Die, 22, 35, 44, 60
evolution, 19–20

Fechner, G. T., psychologist, 36
Fischer, —, geographer, 29

Forster, G., geographer, 4
Forster, R., geographer, 4
Franco-German War, 9
Friedrich, —, geographer, 29

'Geographischer Abend' of Leipzig, 33
Geography
  Ratzel's work in: biogeography 42; human geography, 20, 22–4 42; physical geography, 17, 22, 35; political geography, 36, 38–41
  teaching of, in schools, 34
German minorities in European countries, 37
Germans in U.S.A., 14
Glücksinseln und Träume, 5, 9, 60, 93
Gobineau, J. A., writer and diplomat, 36
Goethe, J. W. von, 41
Götz, W., geographer, 26
Günther, S., mathematical geographer, 26
Guyot, A., geographer, 31

Haeckel, E., zoologist, 7, 19
Hantzch, V., bibliographer of Ratzel, 2, 55–6
Hassert, K., geographer, 5, 30
Heigel, K. von, historian, 26
Helmolt, H., historian, 55
Herder, J. G. von, anthropologist, 24
Hettner, A., geographer, 3, 29
Holtei, K. von, actor and writer, 27
Humboldt, A. von, geographer, 3, 10, 15, 16, 17, 18, 30, 41

Kapp, C., philosopher, 36

INDEX

Kirchhoff, A., geographer, 15
Kittel, R., theologian, 43
Kleine Schriften, 2, 4, 60-1

Lamprecht, K., historian, 38, 43
'Lebensraum, Der', 1, 41, 42, 44, 89
Lowie, R. H., anthropologist, 24, 44

Mackinder, H. J., geographer, 4, 14
Mahan, A. T., historian, 15
map-reading, 28
Martin, C., naturalist, 8
Meer als Quelle der Völkergröße, Das, 1, 15, 60
Merz, J. T., philosopher, 17
migration of animals and peoples, 11, 24, 42
mountain geography, 22

'Nationalitäten und Rassen', 36, 92

Ostwald, W., chemist, 43

Parrella, F., geographer, 38, 39
Penck, A., physical geographer, 31
Perthes, J., publishers, 16, 34
Peschel, O., physical geographer, 26
philosophy, 20, 35
polar exploration, 25
Politische Geographie, 1, 36-41, 60

racial theories, 36
Ratzel, Friedrich
 birth and early years, 5
 schooling, 6
 work as apothecary's assistant, 6, 7
 university education, 7
 zoological work, 7, 8
 travel correspondent for *Kölnische Zeitung*, 8-9, 11-14, 16
 military service, 9
 lecturer and then professor at Munich, 16
 professor at Leipzig, 26
 lectures, 28, 48-9, 49-50
 seminar teaching, 28-9
 marriage and home life, 30, 51
 religious views, 20, 36
 death, 43
Réclus, E., geographer, 15, 35
Richthofen, F. von, physical geographer, 26
Ritter, K., geographer, 3, 10, 17, 19, 20, 31, 38

Sauer, C., geographer, 35
Schäffle, A. E. F., political economist, 1
Schrepfer, H., geographer, 3
sea-power, 15, 36
Sein und Werden der organischen Welt, 7, 57
Semple, E., historian and economist, 23, 30, 31-3
Spencer, Herbert, philosopher, 19, 40, 44
Steinmetzler, J., geographer, ix, 4, 43

Velhagen and Klasing, publishers, 34
Völkerkunde, 1, 22, 23, 24, 25, 58-9

Wagner, H., geographer, 3
Wagner, M., naturalist, 11, 24, 25, 42
Wandertage eines Naturforschers, 13, 15, 57
Wundt, W., philosopher, 43

Zittel, K. von, geologist, 11

FRIEDRICH RATZEL
A BIOGRAPHICAL MEMOIR AND BIBLIOGRAPHY

By HARRIET WANKLYN
(Mrs J. A. STEERS)

Lecturer in Geography in the University of Cambridge

Friedrich Ratzel, born in Baden in 1844, began his career as a pharmacist's apprentice, and took his degree at Heidelberg in the Natural Sciences. Later, he became a journalist, but after travelling widely in Europe and America was appointed a professor of Geography in Munich in 1880. His later academic life was distinguished, and when he died in 1905 his standing as a geographer was high. Since that time his reputation has suffered a setback: he was a prolific and sometimes careless writer, and has been held responsible for the notions of geographical determinism and *Lebensraum*. But a more objective view of his achievements is possible today; and it shows him to be outstanding among the great nineteenth-century German geographers for the range, vigour and courage of his scholarship.

Harriet Wanklyn's short memoir brings together the known details of Ratzel's life in a coherent and accessible form to English readers. The bibliography of Ratzel's works which makes up the second half of the book is an abbreviated version (with English translations of the titles) of the exhaustive bibliography by Viktor Hantzsch. The arrangement in two parts (the first listing complete books and the second contributions to journals and collective works) has been preserved, and only the more ephemeral entries have been omitted.